SpringerBriefs in Political Science

SpringerBriefs present concise summaries of cutting-edge research and practical applications across a wide spectrum of fields. Featuring compact volumes of 50 to 125 pages, the series covers a range of content from professional to academic. Typical topics might include:

- A timely report of state-of-the art analytical techniques
- A bridge between new research results, as published in journal articles, and a contextual literature review
- A snapshot of a hot or emerging topic
- An in-depth case study or clinical example
- A presentation of core concepts that students must understand in order to make independent contributions

SpringerBriefs in Political Science showcase emerging theory, empirical research, and practical application in political science, policy studies, political economy, public administration, political philosophy, international relations, and related fields, from a global author community.

SpringerBriefs are characterized by fast, global electronic dissemination, standard publishing contracts, standardized manuscript preparation and formatting guidelines, and expedited production schedules.

Olivier Serrat

Advancing Environmental Peacebuilding

Pathways for Progress

Springer

Olivier Serrat
School of Continuing Studies
Georgetown University
Washington, DC, USA

ISSN 2191-5466 ISSN 2191-5474 (electronic)
SpringerBriefs in Political Science
ISBN 978-981-92-4027-2 ISBN 978-981-92-4028-9 (eBook)
https://doi.org/10.1007/978-981-92-4028-9

This Springer imprint is published by the registered company Springer Nature Singapore Pte Ltd.
The registered company address is: 152 Beach Road, #21-01/04 Gateway East, Singapore 189721, Singapore

Preface

Throughout history, communities have developed practices that linked environmental stewardship with social stability. While these practices were often localized and implicit, they reflected a long-standing recognition of the interdependence between ecological conditions and social order. In the contemporary era, the scale and complexity of environmental disruption require that these relationships be more explicitly articulated and systematically addressed. As resource pressures intensify, ecosystems degrade, and climate change accelerates, environmental peacebuilding has gained prominence for its potential to illuminate how environmental factors shape the dynamics of peace and conflict. This growing attention underscores the need for approaches that engage directly with the interconnections between ecological systems and human security. Yet despite the field's expansion, its conceptual foundations remain dispersed across disciplines, and its evaluative tools are not fully equipped to address the socio-ecological complexity in which interventions occur.

This SpringerBrief responds to these challenges. It offers a concise, interdisciplinary introduction to environmental peacebuilding and positions evaluation as central to the field's continued development. It traces the historical emergence of environmental peacebuilding, synthesizes contributions from key disciplines, and proposes context-sensitive approaches for assessing interventions in conflict-affected or environmentally fragile settings. It argues that without evaluative frameworks capable of addressing complexity, uncertainty, and the hybrid nature of socio-ecological systems, the field risks falling short of its potential as a pathway to peace. Recent scholarship has highlighted the importance of such integrative approaches, particularly in relation to cooperative environmental governance and climate-linked risks (Serrat, 2025).

This Brief is organized as a series of précis, each addressing a core dimension of the field. The chapters follow a grounded scholarly progression—moving from historical foundations to disciplinary perspectives, then to methodological development and applied illustration—to reflect both the field's current realities and its forward-looking possibilities. Chapter 1 outlines the historical evolution of environmental peacebuilding. Chapters 2 and 3 examine disciplinary contributions from anthropology, economics, environmental science, history, international relations, and sociology. Chapter 4 introduces liberal studies as a means of bridging these domains and strengthening evaluative reasoning. Chapter 5 discusses how ideas develop into scholarly contributions, using the formulation of evaluative criteria as an illustrative case. Chapters 6 and 7 integrate these insights by proposing extensions to the evaluation framework of the Organisation for Economic Co-operation and Development's Development Assistance Committee and presenting new, context-sensitive evaluative criteria for assessing environmental peacebuilding initiatives. Chapter 8 examines a formative effort—the Tonle Sap Initiative—that embodied environmental peacebuilding principles before the field was formally defined, illustrating how integrated environmental governance can serve as peace infrastructure. It offers a rare insight into the genesis of the initiative, back in 2001, providing a window into the formative thinking that anticipated many of the principles now central to environmental peacebuilding. Chapter 9 synthesizes this Brief's central themes and looks ahead to the field's future. It reflects on how environmental peacebuilding must adapt in the Anthropocene, emphasizing a shift from cooperation to co-creation, the need for adaptive and context-sensitive evaluation, and the ethical and intergenerational stakes that will shape emerging practice. In doing so, it positions environmental peacebuilding as essential architecture for building shared futures in an increasingly interdependent world.

This Brief is intended for policymakers, practitioners, and scholars seeking a clear and integrative overview of the field. It aims to support more rigorous analysis, more effective intervention design, and more context-attuned evaluation. Its concise format and interdisciplinary scope also make it suitable for upper-level undergraduate and graduate courses across development, environment, peace, and security studies.

Environmental peacebuilding encourages renewed consideration of how cooperation over shared environments can contribute to stability and the pursuit of lasting peace. By clarifying the field's foundations while pointing toward new evaluative and conceptual pathways, this Brief seeks to contribute to a field that is still taking shape—one whose promise depends on both analytical rigor and sustained imagination. In doing so, it aims to strengthen ongoing efforts to advance scholarship and, ideally, the practice that must continue to evolve.

Washington, DC, USA Olivier Serrat

Reference

Serrat, O. (2025). *Anthropogenic solutions for climate change: Achieving environmental peace*. Springer.

Letter to the Reader

I did not set out to write this SpringerBrief with a specific theory in mind. It developed from a realization that had become clearer over time: the environment is not merely a backdrop to human conflict but frequently the first to absorb the consequences of our divisions and the extractive practices that both shape and result from them, with repercussions that return in ways we rarely anticipate.

That recognition—born less of insight than of accumulated unease—carried through years of engagement across Asia and the Pacific on the practical challenges of environmental management. The unease came from seeing how often well-intentioned interventions faltered for reasons that had little to do with ecology and everything to do with people. Whether working across grassland and wetland ecosystems, extensive livestock and rangeland systems, inland and marine fisheries, land resource evaluation, flood control and irrigation, watershed stewardship, basin-level planning, or the institutional arrangements meant to coordinate these sectors, I saw technical solutions falter—if they did not make matters worse—whenever interests were not shared and trust was thin. I grew dissatisfied with how ecological issues were typically framed as problems to be solved rather than as elements of living systems whose life-bearing interactions require sustained attention.

Everything shifted in 2001 during the effort that came to be called the Tonle Sap Initiative—a chapter revisited at the end of this Brief. In Cambodia, communities were navigating intertwined pressures: rapid population growth following the fall of the Khmer Rouge and the end of decades of civil conflict, ecological stress signals in the Tonle Sap lake–floodplain system, overlapping and still-rebuilding institutional mandates, and livelihood insecurity. The Initiative's approach—grounded in participatory governance, multilevel coordination, and the balancing of conservation with human need—illustrated something fundamental. Cooperation does not arise from declarations or negotiations. It grows from joint engagement with shared environmental concerns and from the slow, patient rebuilding of ties across divides. The Initiative was never framed as "peacebuilding," yet it demonstrated—quietly but unmistakably—how environmental collaboration can generate the conditions for peace long before the term existed. "Everything has been thought of before, but the problem is to think of it again," as Johann Wolfgang von Goethe once observed apropos originality.

This Brief is rooted in those experiences. It is written for readers who recognize that restoring ecosystems and mending social bonds are closely connected tasks. It follows the thread of how environmental collaboration has shaped peace, and it proposes practical improvements for strengthening this work today. My hope is that these pages offer not only analysis, but also a reminder that peace is most reliably supported in the everyday practice of tending the places we share. If this work points toward anything, it is toward choosing care over depletion. Be for the glory, not the degradation. What can you contribute?

Washington, DC, USA Olivier Serrat

Competing Interests The author has no competing interests to declare that are relevant to the content of this manuscript.

Contents

List of Figures

List of Tables

Chapter 1
Roots of Reconciliation: The Emergence of Environmental Peacebuilding

Abstract Environmental peacebuilding coalesced through the long-term convergence of early resource-sharing practices, rising global concern over environmental degradation, and the institutionalization of environmental issues within the United Nations. This précis traces that evolution, showing how global environmental politics, institutional reforms, and interdisciplinary scholarship transformed scattered practices into a distinct and coherent paradigm.

1.1 Building Peace Through Nature: From Practice to Paradigm

The dynamics we now call environmental peacebuilding were shaping realities long before anyone gave them a name. The language is new; the underlying intuition is not. Long before academics codified the idea, communities were coordinating grazing, negotiating water use, rehabilitating degraded landscapes, and stewarding forests in ways that eased tensions, preserved shared resources, and enabled coexistence (Agrawal, 2001; Ostrom, 1990). As the section on early pathways will show, these practices formed the historical bedrock of the contemporary field.

On their own, however, these practices did not create a coherent field. Environmental peacebuilding emerged as a domain only when several historical forces converged. Rising alarm over environmental degradation in the 1960s and 1970s, the surge of global environmental governance following the 1972 Stockholm Conference, and the birth of climate diplomacy in the late 1980s and early 1990s—ultimately leading to the 1992 United Nations Framework Convention on Climate Change (UNFCCC)—collectively propelled ecological concerns to the center of international politics. At the same time, debates on environmental security (Barnett, 2001; Brauch, 2008), post-Cold War peacebuilding (Paris, 2004), and sustainable development underscored how environmental stress could intensify instability or create openings for cooperation. Together, these developments provided the conceptual vocabulary, institutional architecture, and political momentum that crystallized the field.

O. Serrat, *Advancing Environmental Peacebuilding*, SpringerBriefs in Political Science, https://doi.org/10.1007/978-981-92-4028-9_1

A major catalyst was the expanding role of the United Nations Environment Programme (UNEP), founded in 1972 in the aftermath of the Stockholm Conference. Beginning in the late 1990s, UNEP's post-conflict environmental assessments and technical missions made clear that environmental degradation was not just a development concern but a decisive factor in peace and post-conflict recovery (UNEP, 2009). Through this work, UNEP forged an explicit link between environmental management and peacebuilding, showing that ecological restoration could function as a strategic instrument of stabilization. These efforts produced the first systematic body of evidence demonstrating how environmental management contributes to stabilization and durable peace (Bruch et al., 2016).

Parallel advances in research bolstered this institutional shift. Beginning in the early 2000s, interdisciplinary scholarship spanning environmental science, international relations, peace studies, and political ecology clarified concepts, methodological tools, and empirical findings (Matthew et al., 2002; Dresse et al., 2019). By the 2010s, policy, practice, and scholarship had converged into a recognizable domain of theory and action.

Today, environmental peacebuilding refers to the policies, practices, and research that examine how environmental conditions and natural resource governance bear on conflict dynamics, enable cooperation, and support post-conflict recovery (Bruch et al., 2016; Conca & Dabelko, 2002; Dresse et al., 2019). This précis traces the field's historical development—from early resource-sharing arrangements to UNEP's leadership and current climate-security debates—to show how diverse experiences and ideas formed a coherent paradigm. Understanding this evolution, illustrated in Fig. 1.1, is essential for interpreting current approaches and assessing future possibilities in a time of accelerating environmental change.

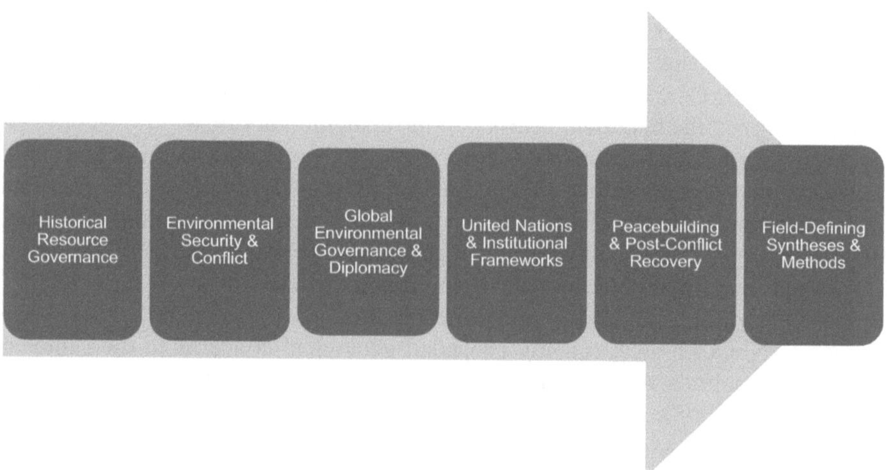

Fig. 1.1 Historical Trajectory of Environmental Peacebuilding. *Note*. This figure synthesizes the historical currents that informed Environmental Peacebuilding. It traces the field's evolution from longstanding traditions of local resource management, through the rise of environmental security debates and the expansion of global environmental governance and diplomacy, to the institutionalization of environmental considerations in peacebuilding and post-conflict recovery—culminating in the more integrative approaches that characterize the field today

1.2 The Long Arc of Environmental Cooperation

Environmental peacebuilding did not begin with academic frameworks, climate diplomacy, or United Nations mandates; it is rooted in the earliest moments when human survival depended on negotiating access to shared land, water, and ecosystems. Long before the term existed, communities recognized that the environment could fracture societies or forge cooperation (Mithen, 2012), and this deep history makes clear that environmental governance has always been inseparable from social stability. Although often portrayed today as a response to climate pressures, post-conflict recovery, and resource scarcity, the field's modern form took shape gradually. A historical perspective reveals not only continuity across eras and diverse regions but also the recurring mechanisms—shared norms, negotiated access, and collective stewardship—that link, say, ancient irrigation compacts to contemporary peacekeeping. The period from the 1980s to the early 2000s was especially decisive, as environmental security debates (Barnett, 2001; Brauch, 2008; Homer-Dixon, 1999; Kahl, 2006), sustainable development agendas, and post-Cold War peacebuilding efforts supplied the conceptual and institutional scaffolding that underpins the field today.

1.2.1 Seeing the Field Historically

Early efforts to manage shared natural resources—often as common-pool systems—were not labeled "peacebuilding," yet they demonstrated how environmental interdependence can encourage dialogue and restraint (Agrawal, 2001; Ostrom, 1990). Tracing these antecedents brings to light the conceptual threads linking past and present approaches.

1.2.2 How Multiple Traditions Converged

Environmental peacebuilding emerged at the intersection of development studies, peace studies, and environmental security, each contributing key insights. Development studies, grounded in post-war reconstruction and decolonization, emphasized sustainable resource management and the governance of shared natural assets (Matthew et al., 2002). Peace studies, which took shape in the 1960s, examined the political and social dimensions of conflict transformation (Lederach, 1997). Environmental security scholarship later emphasized the risks posed by ecological degradation and resource stress (Barnett, 2001; Homer-Dixon, 1999). By the 1990s, these strands were increasingly converging, producing a more comprehensive understanding of how environmental factors shape conflict and cooperation. A historical view shows how this interdisciplinary synthesis evolved through ongoing debate, dialogue, and empirical research (Dresse et al., 2019).

1.2.3 When the Planet Became a Security Issue

As concerns about biodiversity loss, climate change, and competition over natural resources intensified in the late twentieth century, policymakers, practitioners, and scholars increasingly treated environmental issues not only as ecological challenges but also as questions of security and stability (Adger et al., 2014; Gleditsch, 2012). This shift expanded peacebuilding's scope, prompting institutions to consider how environmental stressors might heighten tensions or hinder recovery. By the early 2000s, climate-related security risks had become a recurring theme in international policy debates (Mach et al., 2019), further mainstreaming environmental issues into peace and security agendas. Situating the field within this broader shift highlights its growing responsiveness to changing ecological and geopolitical dynamics (Serrat, 2025a).

1.2.4 Institutionalization: Building on Precedent and Practice

The institutionalization of this work—particularly within the United Nations system—grew out of earlier experiences in humanitarian response, peacekeeping, and post-conflict reconstruction (Paris, 2004). Incorporating natural resource management into peace agreements and folding environmental assessments into recovery planning created early models that later initiatives built upon (UNDP, 2013; UNEP, 2009). UNEP's post-conflict environmental assessments beginning in 1999, the expansion of the United Nations' peacebuilding architecture in 2005, and the mainstreaming of natural resource governance throughout the 2000s reveal how environmental considerations became increasingly embedded in institutional mandates and practice (Bruch et al., 2016). These developments illustrate a field that evolved gradually and adaptively, with early lessons informing later policy design and operational innovation.

1.2.5 Why History Matters for Today's Practice

Across the field's deep-rooted antecedents, recurring patterns stand out—the centrality of local participation, the limits of technocratic approaches, and the trust-building potential of environmental cooperation—all of which remain relevant across diverse contexts (Autesserre, 2014; Pretty & Ward, 2001). Insights accumulated from the 1970s through the 2010s point to the need for adaptive strategies that can respond to shifting environmental and political conditions, reinforcing the value of historical perspective for more effective practice.

1.3 Early Pathways: Natural Resource Cooperation Before Environmental Peacebuilding

Environmental peacebuilding's earliest expressions grew from the pragmatic need to manage shared ecological systems in ways that reduced tensions, supported coexistence, and sustained livelihoods. Long before scholars articulated environmental security or practitioners designed formal peacebuilding interventions, communities were developing mechanisms to govern common resources (Agrawal, 2001; Ostrom, 1990). Although rarely described as "peacebuilding," these arrangements performed many of the functions associated with contemporary practice. This section surveys ten such cases to illustrate the variety and sophistication of these early approaches.

1.3.1 Sacred Landscapes and Ritual Stewardship

In many cultural traditions, collective management formed around sacred landscapes—forests, mountains, pilgrimage routes, and water bodies imbued with spiritual significance. These areas were often safeguarded through shared norms rather than formal treaties. Among many examples are Akan sacred groves in Ghana protected for more than 500 years, Hopi and Navajo stewardship of ancestral mesas dating to the twelfth century, and South Asian temple tanks maintained since the Chola period (ninth–thirteenth centuries). These instances attest that collective stewardship was not only material but also cultural and symbolic, reinforcing shared identity and mutual respect (Ali, 2007; Posey, 1999; Sheridan & Nyamweru, 2008).

1.3.2 Agricultural Commons and Communal Land-Use Systems

Agricultural landscapes required collaboration among households, extended kin groups, and villages whose livelihoods depended on shared fields, soils, and seasonal rhythms. Medieval European open-field systems documented from the ninth century and lasting into the eighteenth century coordinated crop rotations, fallowing cycles, plowing schedules, and post-harvest grazing, reducing disputes by aligning land-use decisions across entire communities. In Southeast Asia, sixteenth-century Tai chronicles describe rotational Swidden agreements that regulated cycles and prevented encroachment, while in West Africa, communal tenure systems recorded in nineteenth-century ethnographies—but rooted in much older oral traditions—governed planting rights and soil fertility management. These systems were more than agricultural practices: they were governance structures that synchronized expectations, embedded environmental cooperation into daily life, and minimized conflict (Netting, 1993; Ostrom, 1990).

1.3.3 Wildlife and Game Management

Coordinated efforts around wildlife preceded modern conservation by thousands of years. Many societies developed shared rules for managing access to high-value species, hunting grounds, and migration corridors. In East Africa, pre-colonial Maasai agreements regulating elephant corridors and hunting rights—referenced in nineteenth-century accounts—helped prevent conflict. In Europe, the 1424 Scottish Statute of the Forest governed shared hunting forests, while Indigenous nations across the North American Plains negotiated 18th- and early nineteenth-century bison-hunting protocols to prevent clashes between mobile groups. These systems recognized wildlife as a mobile, transboundary resource whose mismanagement could provoke violence. By establishing norms for harvest, movement, and territorial use, wildlife governance served as a stabilizing mechanism that balanced ecological pressures with social cohesion (Berkes, 1989; Niamir-Fuller, 1999).

1.3.4 Pastoral Mobility and Grazing Systems

In arid and semi-arid regions, pastoralist groups developed intricate systems for managing grazing lands and water points. These arrangements governed seasonal mobility, negotiated access to pastures, and articulated protocols for resolving disputes. Examples from Central Asia, the Horn of Africa, and the Sahel—many of which can be traced back several centuries—show how pastoral mobility systems have long functioned as peacebuilding institutions by preventing competition from escalating into violence. The Somali *xeer* (customary law) system, with elements traceable to the seventh century, the Kazakh and Kyrgyz transhumance routes mapped in eighteenth-century Russian surveys, and Tuareg agreements on dry-season wells described in nineteenth-century travel accounts stress the importance of social institutions in managing environmental interdependence (Behnke et al., 1993).

1.3.5 Forest and Timber Governance

Forests were another domain in which cooperation arose early and often. Shared forest systems required coordination to manage hunting rights, regulate timber extraction, and prevent fires. Alpine inter-village forest compacts such as the 1370 Valais Forest Charter, Scandinavian timber agreements linked to sixteenth-century naval expansion under Denmark–Norway, and Joseon-era (1392–1910) protocols for managing Korean border forests all reflected an understanding that forests were strategic resources. By establishing rules, monitoring mechanisms, and shared responsibilities, forest governance served as a stabilizing force across regions.

Contemporary initiatives such as the Heart of Borneo show how transboundary forest cooperation can foster dialogue, shared stewardship, and ecological resilience (Serrat, 2025b).

1.3.6 Disaster-Risk Cooperation

Environmental hazards—droughts, earthquakes, floods, and storms—have long compelled communities to collaborate despite their differences. Across the Mediterranean, coastal settlements rebuilt shared sea walls after the 1573 Adriatic storm and maintained joint early-warning practices. Chinese drought-relief systems, including Qing-era (1644–1912) communal granaries, helped avert famine-driven conflict. In South Asia, eighteenth-century Mughal records describe inter-village embankment maintenance and flood-warning networks that reduced the risk of catastrophe. These forms of collective action lowered the likelihood that environmental shocks would trigger social unrest or inter-group violence. Disaster-risk cooperation illustrates how shared vulnerability can generate durable collective action, reinforcing interdependence and trust (Bankoff, 2003; Wisner et al., 2004).

1.3.7 Shared Infrastructure and Environmental Engineering

Many societies cooperated to build and maintain shared environmental infrastructure (e.g., irrigation canals, drainage networks, flood-control embankments, terracing systems). These projects demanded long-term planning, collective labor, and coordination across communities. Pharaonic Nile flood-basin management dating to ca. 3000 BCE, the Inca terrace systems at Moray and Pisac (ca. 1400–1530), and Bali's *subak* irrigation cooperatives—documented since the eleventh century—make obvious how environmental engineering fostered joint action by creating shared responsibilities and mutual benefits. They also show that environmental management was inseparable from broader political and social institutions (Mithen, 2012).

1.3.8 Fisheries and Marine Resource Agreements

Coastal and riverine communities frequently cooperated to manage fisheries and marine ecosystems. Migratory fish stocks, spawning grounds, and shared coastal zones required coordination to avoid overexploitation. North Sea herring treaties from the 1320s, Japanese–Korean fishery agreements negotiated between 1609 and 1611, and customary Pacific Northwest Indigenous fishing compacts documented in nineteenth-century ethnographies all illustrate how communities developed norms and institutions to regulate access. These systems often included gear restrictions, seasonal closures, and conflict-resolution mechanisms, demonstrating that marine resource governance was a highly sophisticated and cooperative endeavor long before modern environmental law (Berkes, 1985).

1.3.9 Post-Conflict Land Restoration

War scars landscapes as deeply as it scars the people and institutions that endure it. Its environmental impacts range from manageable degradation to severe destruction: abandoned fields, contaminated water sources, exhausted soils, shattered irrigation systems, and terrain rendered dangerous by fortifications or unexploded ordnance. In the aftermath, land restoration becomes essential to rebuilding livelihoods. Examples include post-World War II demining and agricultural rehabilitation in Europe (1945–1955); post-independence restoration in Mozambique and Uganda (1960s–1990s); and post-Vietnam War agricultural recovery in the late 1970s and 1980s. Similar efforts followed the 1996 peace accords in Guatemala, the 1996–2006 Maoist conflict in Nepal, the scorched-earth destruction of 1999 in Timor-Leste, and the post-Khmer Rouge period in Cambodia, where communities reclaimed farmland, repaired irrigation systems, and rebuilt rice production. Although seldom labeled as peacebuilding, post-conflict land restoration has long functioned as a vital means of restoring livelihoods, rebuilding social trust, and reducing the risk of renewed violence (Unruh, 2003; Unruh & Williams, 2013).

1.3.10 Transboundary Water Management

Water has long been a catalyst for cooperation. Because aquifers, lakes, and rivers routinely cross political boundaries, communities and states have had to negotiate access, distribution, and quality. Mesopotamian irrigation agreements (ca. 2500–2000 BCE), medieval Rhine River compacts (twelfth–fourteenth centuries), and the 1909 Boundary Waters Treaty between Canada and the United States—which created the International Joint Commission—show how water governance structured interactions between actors who might otherwise have remained in conflict. The 1957 Mekong Committee, one of the earliest modern multilateral river institutions, reflects mid-twentieth-century approaches to shared water management. Contemporary initiatives, including twenty-first-century resource-based cooperation in the Mekong River Basin, demonstrate how ecological interdependence continues to create space for dialogue and joint governance (Dinar et al., 2007; Serrat, 2025c; Wolf, 2007; Zeitoun & Mirumachi, 2008).

1.3.11 What Early Pathways Reveal

Several themes stand out. First, cooperation was driven by necessity rather than idealism: shared ecological dependence compelled actors to negotiate, coordinate, and coexist. Second, cultural and spiritual values shaped environmental management alongside material needs, reinforcing norms of restraint, reciprocity, and shared responsibility (Posey, 1999). Third, cooperation operated across multiple levels, from local arrangements to regional systems and, at times, inter-state governance. Fourth, as management expanded from local practices to broader regional systems, institutions became essential for governing shared resources and preventing disputes (Agrawal, 2001; Ostrom, 1990). Fifth, under acute pressures, domains such as disaster-risk cooperation and post-conflict land restoration show how collaboration forged under stress supported livelihoods, rebuilt trust, and reduced the risk of renewed violence (Unruh, 2003).

These early pathways formed the foundations of contemporary environmental peacebuilding. This long arc of cooperation now finds a modern analogue in global climate governance, which—like earlier forms of shared resource management—arises from ecological interdependence and requires coordinated action across political boundaries. In this sense, climate governance functions as an eleventh pathway, extending longstanding traditions of environmental cooperation to a planetary scale. Figure 1.2 distills these pathways into a comparative framework showing the ten early forms of cooperation, the scales at which they operated, and their continued relevance in broader or more formalized systems.

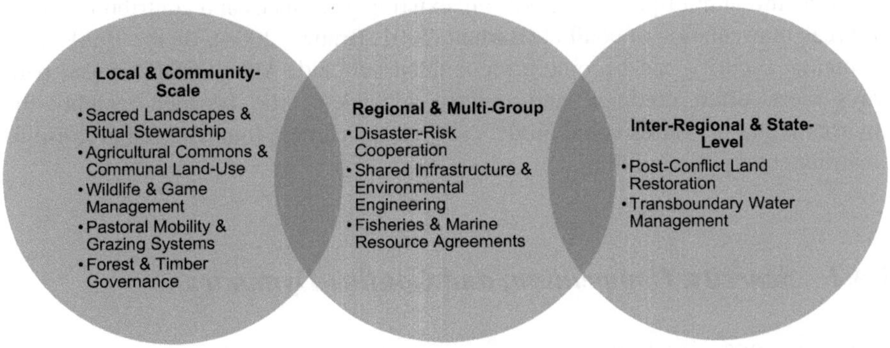

Fig. 1.2 Ten Early Pathways of Environmental Cooperation. Note. These categories summarize forms of environmental cooperation found across societies. They vary in scale and historical depth, with some rooted in longstanding communal practices and others transformed more recently by formal institutions and state systems. The typology is intended as an analytical guide, not an exhaustive catalogue

1.4 Environmental Stress as a Catalyst for Conflict and Cooperation

Environmental peacebuilding arose in part from a growing recognition that environmental factors influence both conflict and cooperation (Homer-Dixon, 1999; Kahl, 2006). While early pathways shed light on how shared resource management could reduce tensions, the late twentieth century introduced a different kind of insight—one focused on the destabilizing effects of ecological decline. This section examines how resource-related strains became central to debates about security and peacebuilding, and how this shift helped solidify the field's analytical underpinnings.

1.4.1 From Environmental Degradation to Security Concerns

Deforestation, land degradation, water scarcity, and declining agricultural productivity have been tied to displacement, livelihood insecurity, and competition over dwindling resources (Homer-Dixon, 1999). As policymakers, practitioners, and scholars observed these correlations between ecological decline and social unrest—particularly from the 1980s onward—the idea that environmental stress could drive conflict gained prominence. These insights challenged traditional security paradigms by demonstrating that threats to stability could emerge not only from political or military sources but also from intensifying environmental pressures.

This shift was especially noticeable in the emergence of environmental security as a distinct discourse. Analysts argued that environmental degradation could function as a "threat multiplier," intensifying existing grievances and contributing to the outbreak or escalation of conflict (Barnett, 2001; Brauch, 2008). By the 1990s, environmental security had become a recognized subfield. Although the causal pathways were often complex and context-dependent, the broader insight—that environmental pressures could erode stability—became a foundational premise for environmental peacebuilding.

1.4.2 Scarcity, Competition, and Conflict Dynamics

In the late 1980s and 1990s, resource scarcity became one of the most widely discussed mechanisms linking environmental stress to conflict. In regions where communities depend heavily on natural resources for their livelihoods, declining availability of arable land, grazing areas, or water can heighten competition and strain social relations (Homer-Dixon, 1999; Kahl, 2006). Historical and contemporary examples—from the Sahel's drought cycles in the 1970s and 1980s to mounting land-use pressures in the Horn of Africa during the 1990s and 2000s—show

how environmental stress can ignite localized violence, particularly when compounded by economic marginalization, social fragmentation, or weak governance.

However, the relationship between resource pressures, competition, and conflict is not deterministic. Environmental pressures do not inevitably lead to violence; rather, they interact with economic, political, and social factors in ways that can either exacerbate tensions or encourage cooperation. In many cases, governance failures—not scarcity alone—determined whether tensions escalated (Le Billon, 2005). Recognizing this complexity is essential for understanding why environmental peacebuilding emphasizes dialogue, governance, and institutions rather than solely technical solutions.

1.4.3 Environmental Pressure as a Driver of Collaboration

While resource-related strains can contribute to conflict, they can also create incentives for cooperation. Shared vulnerability to drought, flooding, or resource depletion may encourage communities and states to collaborate on joint solutions. Examples from the 1980s and 1990s—such as cooperative water management in the Middle East and regional drought responses in southern Africa—show how environmental challenges have prompted the creation of new communication channels, institutions, or agreements that help build trust and reduce the risk of conflict (Dinar et al., 2007; Wolf, 2007).

This dual potential—ecological stressors functioning as both risks and openings—became a defining feature of environmental peacebuilding. It underscored the need for strategies that address not only the ecological dimensions of stress but also the economic, political, and social structures that influence how societies ultimately respond.

1.4.4 Climate Change and the Broadening of Environmental Peacebuilding

Shifting climate patterns intensified discussions about ecological pressures by introducing new forms of risk and uncertainty. As extreme weather events, rising temperatures, and shifting rainfall patterns began to affect livelihoods and ecosystems—trends documented by the Intergovernmental Panel on Climate Change (IPCC) beginning with its First Assessment Report in 1990—policymakers, practitioners, and scholars came to see climate impacts as potential drivers of instability (Adger et al., 2014; Gleditsch, 2012; Mach et al., 2019). This widened the scope of environmental peacebuilding by highlighting the need for adaptive governance, long-term planning, and greater resilience building.

Climate-related vulnerabilities also reinforced the importance of cooperation. Transboundary climate impacts—such as regional drought patterns or shared river basin variability—showed clearly that no single actor could address these challenges alone. By the 2000s, this recognition helped legitimize environmental peacebuilding as a field concerned not only with post-conflict recovery but also with preventing conflict through collaborative decision-making systems (Busby, 2022).

1.4.5 How Environmental Stress Shaped the Field's Analytical Core

The growing attention to resource-related strains contributed to the development of analytical frameworks that sought to explain how environmental factors influence conflict and cooperation. During the 1990s and early 2000s, these frameworks emphasized the importance of governance, institutions, and social cohesion in mediating the effects of environmental pressures (Dresse et al., 2019; Ide, 2019). They also underscored the need for cross-disciplinary interpretive lenses that combine ecological, political, and socio-economic perspectives (Matthew et al., 2002).

1.5 Global Environmental Movements and the Rise of Climate Diplomacy

The emergence of environmental peacebuilding cannot be understood without examining the global environmental movements and diplomatic processes that transformed international awareness of ecological issues in the late twentieth century. These movements expanded the intellectual and political room for linking environmental challenges to wider discussions about peace, security, and development (Biermann & Pattberg, 2008; Sachs, 1993). This section traces the evolution of global environmental consciousness, the rise of environmental diplomacy, and the ways in which these developments helped anchor environmental peacebuilding as a field of inquiry and practice.

1.5.1 Awakening to a Shared Planet: The Making of Modern Environmental Awareness

The modern environmental movement gained momentum in the 1960s and 1970s, propelled by growing public concern about biodiversity loss and the long-term consequences of industrialization including pollution. Influential publications—such as Rachel Carson's *Silent Spring* (1962)—along with grassroots activism and scientific warnings converged to create a sense of global ecological vulnerability. This shift in public consciousness laid the groundwork for international cooperation by framing environmental degradation as a shared challenge that transcended national boundaries.

Stockholm (1972) marked a pivotal moment in this transformation. By elevating environmental protection to a matter of global concern, the conference created the normative basis for linking ecological well-being to human security (Biermann & Pattberg, 2008). Although the conference did not explicitly address environmental peacebuilding, it introduced the idea that environmental stewardship was integral to stability, thereby opening conceptual space for later attempts to connect environmental oversight with conflict prevention and recovery.

1.5.2 Environmental Governance and Institutional Innovation

Following Stockholm (1972), a series of international agreements and institutions developed to address specific environmental challenges. Treaties on air pollution, biodiversity, hazardous waste, marine pollution, and ozone depletion—many negotiated between the late 1970s and early 1990s—reflected a growing recognition that environmental problems required coordinated global responses (Biermann & Pattberg, 2008).[1] These institutional advances created a dense institutional architecture that underpinned cooperation, information sharing, and joint monitoring across borders.

The establishment of UNEP in 1972 institutionalized environmental governance within the United Nations system. Through assessments, policy guidance, and capacity-building, UNEP helped elevate environmental issues on the global agenda and provided mechanisms for integrating ecological considerations into development and security planning (UNEP, 2009). These institutional innovations laid the groundwork for environmental peacebuilding by demonstrating how environmental stewardship could be woven into broader international policy frameworks (Mitchell, 2003).

[1] Following the 1972 Stockholm Conference, a series of major environmental agreements emerged, beginning with the Convention on International Trade in Endangered Species of Wild Fauna and Flora in 1973. In 1979, states adopted both the Convention on Long-Range Transboundary Air Pollution, the first treaty to address cross-border air pollution, and the Convention on the Conservation of Migratory Species of Wild Animals, known as the Bonn Convention. The 1980s brought further institutionalization through the United Nations Convention on the Law of the Sea in 1982, the Vienna Convention for the Protection of the Ozone Layer in 1985, and the Montreal Protocol on Substances that Deplete the Ozone Layer in 1987. By 1989, the Basel Convention on the Control of Transboundary Movements of Hazardous Wastes and Their Disposal sought to regulate hazardous-waste trade. In the early 1990s, the Convention on Environmental Impact Assessment in a Transboundary Context appeared in 1991, followed in 1992 by three major outcomes of the Rio Earth Summit: the Convention on Biological Diversity, the UNFCCC, and the launch of Agenda 21 and the Commission on Sustainable Development.

1.5.3 Climate Diplomacy and the Expansion of Environmental Politics

Climate diplomacy was established as a central pillar of the global environmental policy regime in the late twentieth century. Scientific consensus on climate change, articulated through assessments by the IPCC beginning in 1990, underscored the urgency of collective action (Adger et al., 2014). The 1992 UNFCCC and subsequent agreements such as the 1997 Kyoto Protocol and the 2015 Paris Agreement signaled the international community's commitment to addressing climate risks.

As climate impacts became more visible, climate diplomacy expanded the political relevance of environmental issues by linking them to economic development, energy policy, and national security. Policymakers increasingly recognized that climate-related stress could exacerbate social tensions and undermine livelihoods (Busby, 2022; Gleditsch, 2012). This recognition helped link environmental regulation more explicitly to peacebuilding, reinforcing the idea that managing environmental risks was integral to maintaining stability.

1.5.4 Environmental Movements and the Politics of Solidarity

Global environmental movements played a crucial role in shaping the discourse that would inform environmental peacebuilding. Transnational networks of activists, scientists, and civil society organizations—particularly active from the 1980s onward—advocated for environmental justice, sustainable development, and the rights of marginalized communities (Sachs, 1993). Their advocacy drew attention to the social dimensions of environmental degradation and emphasized the need for inclusive, participatory approaches to environmental stewardship.

These movements reframed environmental issues as inherently political, rooted in questions of equity, access, and power. This reframing aligned environmental governance with the core concerns of peacebuilding: equity, participation, and the transformation of structural grievances. By foregrounding justice and participation, environmental movements helped articulate the moral foundation of environmental peacebuilding (Matthew et al., 2002).

1.5.5 Linking Environmental Governance to Peace and Security

The convergence of global environmental awareness, institutional innovation, and climate diplomacy created fertile ground for normalizing environmental considerations into peace and security agendas. As environmental stressors became more prominent in international discussions—especially from the 1990s onward—policymakers, practitioners, and scholars began to explore how environmental governance could contribute to conflict prevention, resilience, and post-conflict recovery (Bruch et al., 2016).

This shift accelerated as recognition grew that environmental cooperation can build trust, strengthen institutions, and support long-term stability. By tying environmental management to broader peace and security objectives, these developments helped establish the foundations of environmental peacebuilding as a distinct field.

1.6 Building the Architecture: The United Nations and Emerging Policy Frameworks

The institutionalization of environmental peacebuilding was propelled by the steady elevation of environmental considerations into the mandates, assessments, and operational frameworks of international organizations—most prominently the United Nations. While early cooperative ecosystem management grew organically from local and regional needs, it was the formal recognition of environmental issues within global institutions that helped consolidate environmental peacebuilding as a coherent field (UNDP, 2013; UNEP, 2009). These global developments set the stage for more formal integration within the United Nations system, shaping the policies and practices that continue to influence the field today.

1.6.1 Expanding Environmental Mandates Within the United Nations

The United Nations played a central role in elevating environmental issues within international peace and security discourse. Initially, environmental concerns were addressed primarily through development and humanitarian channels, with limited connection to peacebuilding. However, as evidence accumulated linking environmental degradation to instability—particularly from the 1980s onward—the United Nations began to incorporate environmental considerations into its broader mission (Barnett, 2001; Brauch, 2008).

One early milestone came with the incorporation of environmental assessments into post-conflict recovery planning. Beginning in the late 1990s, these assessments showed both the scale of environmental damage and how degradation could hinder recovery (UNEP, 2009). By documenting the environmental impacts of conflict, the United Nations helped establish that environmental restoration was not merely a development concern but a pillar of sustainable peace. Given the United Nations' influence on global norms and operational practice, its adoption of environmental considerations had a multiplier effect across peacebuilding institutions (Bruch et al., 2016).

1.6.2 *Integrating Environment into Peacekeeping and Post-Conflict Recovery*

As peacekeeping operations expanded in scope and complexity—especially after the end of the Cold War in the 1990s—the United Nations increasingly recognized that environmental factors could influence mission outcomes. Peacekeeping missions often operated in fragile ecosystems, relied on local resources, and interacted with communities whose livelihoods depended on the environment. This recognition led to the development of environmental guidelines for peacekeeping missions in the early 2000s and to measures to integrate environmental considerations into mission planning (UNEP, 2009).

In parallel, post-conflict recovery frameworks began to incorporate natural resource management as a key element of stabilization. Throughout the 2000s, programs to manage forests, rehabilitate agricultural land, and restore water systems were increasingly understood as essential for rebuilding livelihoods and reducing the risk of renewed conflict (UNDP, 2013). These programs revealed how ecological interventions could advance broader peacebuilding objectives by addressing underlying sources of vulnerability (Le Billon, 2005).

1.6.3 *Policy Frameworks Connecting Environment, Conflict, and Peace*

The appearance of policy frameworks explicitly linking environmental issues to conflict and peace marked a significant step in the institutionalization of environmental peacebuilding. Beginning in the late 1990s and accelerating in the 2000s, reports by United Nations agencies, international financial institutions, and research organizations highlighted the role of natural resources in conflict dynamics and emphasized the need for holistic approaches to resource governance (Bruch et al., 2016; Matthew et al., 2002).

These frameworks offered conceptual clarity and operational guidance, showing how environmental management could support conflict prevention, peacekeeping, and post-conflict recovery. They also stressed the importance of governance, participation, and equity—principles that later became central to environmental peacebuilding practice (Dresse et al., 2019).

1.6.4 *The United Nations Environment Programme and Other Specialized Agencies*

Beginning in the early 2000s, UNEP's Post-Conflict and Disaster Management Branch (renamed the Disasters and Conflicts Programme) conducted environmental assessments in conflict-affected regions and provided technical assistance to governments emerging from war (UNEP, 2009). These missions, which expanded throughout the decade, demonstrated the practical relevance of environmental interventions in fragile contexts and generated some of the earliest empirical foundations for environmental peacebuilding (Bruch et al., 2016).

During the same period, other specialized agencies of the United Nations—including the Food and Agriculture Organization (FAO), the United Nations High Commissioner for Refugees (UNHCR), and the United Nations Development Programme (UNDP), established in 1945, 1950, and 1965, respectively—also began to integrate environmental considerations into their post-conflict programming. FAO's work on land tenure and resource governance from the early 2000s, UNDP's integration of sustainable livelihoods and natural resource management into recovery frameworks throughout the 2000s, and UNHCR's environmental efforts in refugee-hosting regions during the mid-2000s underscored that effective environmental management was vital for achieving peace and stability (Unruh, 2003; Unruh & Williams, 2013). These agencies demonstrated that environmental management was not a niche concern but a core, cross-cutting requirement for effective recovery.

1.6.5 Institutionalization Through Practice and Learning

The formalization of environmental peacebuilding was not driven solely by policy frameworks; it also developed through practical experience and ongoing learning. As international organizations—including United Nations agencies, regional bodies, and specialized international nongovernmental organizations—implemented environmental interventions in conflict-affected settings, they gained insights into what worked, what did not, and why. Beginning in the late 1990s and early 2000s, these lessons helped steer subsequent policies and improved practice. Among these actors, UNEP's early fieldwork was especially influential (Bruch et al., 2016; UNEP, 2009).

In Afghanistan, UNEP's 2002–2003 post-conflict environmental assessment documented extensive degradation of agricultural land, forests, and water systems, offering a basis for integrating environmental restoration into recovery efforts (UNEP, 2009). In Liberia, UNEP's 2004–2006 support for post-war reforms in forestry and land governance aimed to reduce the resource-related grievances that had contributed to conflict, showing how environmental institutions could help rebuild state legitimacy (Le Billon, 2005; Unruh, 2003). Beyond individual countries, cooperative arrangements such as the Nile Basin Initiative demonstrated how shared resource governance could support regional dialogue and joint planning (Dinar et al., 2007; Wolf, 2007).

Over time, environmental peacebuilding came to take toot within the operational cultures of many international organizations. By the mid-2000s, it appeared in assessment tools, strategic planning processes, and training programs (UNDP, 2013), ensuring that environmental considerations were treated as integral rather than peripheral. By the 2010s, it had become a recognized component of international peacebuilding practice (Bruch et al., 2016).

1.6.6 Why Institutional Milestones Matter

The institutional milestones traced in this section underline the central role of global governance structures in shaping the evolution of environmental peacebuilding. By integrating environmental considerations into peacekeeping, recovery, and development frameworks—particularly from the late 1990s onward—the United Nations and its partners helped legitimize the field and supplied the mechanisms needed to translate its principles into practice (UNDP, 2013; UNEP, 2009). These shifts also energized academic networks and scholarly frameworks, further consolidating environmental peacebuilding as a distinct domain of inquiry. This consolidation extended beyond scholarly debates, providing the intellectual infrastructure necessary to support effective, evidence-based practice (Dresse et al., 2019).

1.7 Building the Knowledge Base: Foundational Scholarship and Academic Networks

The consolidation of environmental peacebuilding as a distinct field relied on a growing knowledge base nourished by interdisciplinary scholarship and collaborative research networks. This section examines how academia contributed to that evolution, showcasing the foundational studies, methodological advances, and intellectual debates that helped define the field.

1.7.1 Interdisciplinary Foundations

Environmental peacebuilding has, from the very beginning, been shaped by ideas and perspectives drawn from a wide range of disciplines. Anthropology puts the accent on cultural practices, local knowledge systems, and community mechanisms for managing shared resources (Posey, 1999). Development studies revolve around improving people's livelihoods, promoting equity, and building institutional capacity (Pretty & Ward, 2001). Economics explains how incentives, market structures, and distributional effects mediate patterns of resource competition and cooperation. Environmental science provides empirical data on ecological change and resource dynamics (Adger et al., 2014). History traces long-term patterns of adaptation, conflict, and environmental governance (Mithen, 2012). International relations examine how global norms, interstate cooperation, and jointly managed transboundary resources influence the dynamics of security (Biermann & Pattberg, 2008).

Sociology offers perspectives on social cohesion, collective action, and community organization under ecological stress. Political science contributes theories of conflict, cooperation, and governance (Homer-Dixon, 1999; Kahl, 2006).

The interdisciplinary character of environmental peacebuilding is not an academic preference: it reflects the inherent complexity of the issues at stake. Addressing these challenges requires understanding how ecological systems intersect with economic incentives, political institutions, and social norms, among other factors. Scholars recognized early on—particularly from the 1990s onward—that no single discipline could capture these dynamics, prompting the creation of research networks designed to bridge disciplinary boundaries (Dresse et al., 2019).

1.7.2 The Ideas That Shaped the Field

Several strands of scholarship played particularly influential roles in shaping the field. Early work on environmental security—emerging prominently in the late 1980s and early 1990s—focused on the potential for environmental stress to contribute to conflict, while also acknowledging the conditions under which cooperation might come to the fore (Barnett, 2001; Brauch, 2008). Studies of common-pool resource management, especially following Elinor Ostrom's seminal work in 1990, demonstrated how communities could develop effective governance systems to manage shared resources sustainably (Agrawal, 2001; Ostrom, 1990). Research on post-conflict recovery, which expanded significantly in the 1990s and early 2000s, emphasized the importance of natural resource governance in rebuilding livelihoods and preventing renewed violence (Paris, 2004; Unruh, 2003).

These contributions helped establish the intellectual lineage of environmental peacebuilding. They clarified the mechanisms through which environmental factors influence conflict dynamics, identified the governance structures needed to manage shared resources, and highlighted the potential for environmental cooperation to support peacebuilding (Ide, 2019). Over time—particularly throughout the 2000s—academics developed more integrated frameworks that combined these insights, providing a coherent analytical basis for the field (Matthew et al., 2002). These insights provided the conceptual scaffolding for environmental peacebuilding.

1.7.3 The Rise of Research Networks

The growth of environmental peacebuilding was reinforced by the advent of research consortia that connected policymakers, practitioners, and scholars. These networks came together in the late 1990s and early 2000s, forming spaces where researchers could exchange empirical insights, debate theoretical frameworks, and sharpen their methodological approaches. Collaborative projects, conferences, and workshops fostered a sense of community and encouraged cross-disciplinary exchange. Early initiatives such as the Environmental Change and Security Program at the Wilson Center (founded in 1994), the Peace Research Institute Oslo's expanding climate-security collaborations after 1998, and the Environmental Peacebuilding Institute (launched in 2008) provided some of the first sustained spaces for interdisciplinary dialogue. University-based centers—at the University of Maryland, Uppsala University, and ETH Zurich—further anchored the field by grooming scholars and producing foundational research. These networks helped cultivate a shared vocabulary and conceptual baseline for a field that had hitherto been fragmented across multiple disciplines.

These networks also played a crucial role in translating academic insights into policy and practice by creating channels through which evidence could inform programming and institutional reforms. By engaging with civil society groups, governments, and international organizations—especially during the 2000s—scholars helped ensure that environmental peacebuilding initiatives were informed by empirical evidence and grounded in local realities (Matthew et al., 2002). The Environmental Peacebuilding Association (EnPax), formally launched in 2018 but rooted in earlier collaborations dating back to 2010, further institutionalized these exchanges by linking academics with practitioners from UNDP, UNEP, and regional organizations. Through joint fieldwork, shared datasets, and comparative research—many of which accelerated after 2005—these networks reinforced methodological rigor and helped consolidate environmental peacebuilding as a coherent domain of inquiry. This engagement strengthened the field's relevance and contributed to its broader institutionalization.

1.7.4 Methodological Innovations and Empirical Evidence

As the field matured, researchers developed more advanced methodological tools for studying the links between environment, conflict, and cooperation. Advances in climate modeling, community-level reporting tools, geographic information systems, and satellite-based remote sensing—technologies that became widely accessible in the 1990s and 2000s—provided new tools for analyzing environmental change and its social impacts (Adger et al., 2014; Mach et al., 2019). Qualitative approaches such as ethnography and participatory research provided a deeper understanding of community dynamics and local governance practices.

These methodological advances created a stronger empirical foundation. Case studies from diverse regions—many published in the 2000s and early 2010s—showed how environmental interventions could support peacebuilding (Bruch et al., 2016), while comparative analyses clarified the conditions that made such efforts more successful (Rustad & Binningsbø, 2012). This evidence base refined the field's conceptual frameworks and informed the development of more nuanced policy guidelines (Dresse et al., 2019).

1.7.5 Debates and Critical Perspectives

The development of environmental peacebuilding was accompanied by debates over its assumptions, implications, and methods. Some scholars—particularly in the late 1990s and 2000s—questioned causal claims linking environmental pressures to conflict, arguing that economic and political factors often played larger roles (Gleditsch, 2012). Others warned against technocratic approaches that ignored equity, local agency, and power (Autesserre, 2014). These debates pushed the field away from deterministic models and toward perspectives that foreground agency, context, and governance, fostering more sophisticated analyses and a more holistic, context-sensitive approach (Barnett, 2001; Brauch, 2008).

1.7.6 Consolidating a Field

As practical experience accumulated, research frameworks expanded, and debates sharpened assumptions, environmental peacebuilding gradually emerged as a distinct domain of inquiry. This consolidation became especially visible in the 2010s, when the field developed its own conceptual frameworks, professional networks, and research agendas, while maintaining strong connections to related fields such as conflict studies, environmental governance, and sustainable development (Dresse et al., 2019).

The consolidation extended beyond well beyond academic circles; it provided the intellectual infrastructure needed to support effective practice. By building a comprehensive knowledge base, researchers helped ensure that environmental peacebuilding undertakings were grounded in empirical evidence, informed by diverse perspectives, and responsive to the complexities of real-world contexts (Conca & Dabelko, 2002; Matthew et al., 2002).

1.8 Synthesis: How Historical Currents Converged into a Distinct Practice

Growing awareness of environmental pressures, the rise of global environmental movements, institutional reforms, the consolidation of scholarly work, and accumulated experience in resource cooperation all contributed to the formation of a coherent field.

1.8.1 Practice Informs Theory, Theory Informs Practice

One of the tenets of environmental peacebuilding is the reciprocal relationship between practice and theory. Early examples of natural resource cooperation—whether in community-based governance, post-conflict land restoration, or transboundary water management—proved that environmental management could support stability (Ostrom, 1990; Unruh, 2003; Wolf, 2007). Documented largely in the mid- to late twentieth century, these cases predated formal theorization but later became foundational evidence for policymakers, practitioners, and scholars seeking to understand how environmental factors shape peace and conflict.

As academic frameworks developed—especially during the 1990s and 2000s—they clarified why some environmental interventions succeeded while others faltered, such as why locally grounded water-sharing arrangements fostered cooperation while externally imposed land-use reforms often provoked resistance (Homer-Dixon, 1999; Kahl, 2006). This reciprocal relationship between practice and theory kept the field anchored in real-world conditions (Ide, 2019).

1.8.2 Governing Across Scales: Local to Global

Environmental peacebuilding also emerged through the integration of governance systems across local, national, and international scales. Local communities contributed crucial insights into resource-management practices that deepened cooperation and enhanced collective resilience (Posey, 1999; Sheridan & Nyamweru, 2008). National governments drafted policies on environmental regulation—such as land tenure and water allocation—particularly as environmental governance reforms expanded in the 1990s and 2000s (Le Billon, 2005). International organizations provided frameworks, legitimacy, and resources for environmental interventions in conflict-affected settings (UNDP, 2013; UNEP, 2009).

This multiscalar integration enabled environmental peacebuilding to address complex challenges that transcend political boundaries. It also underscored that cooperation over natural resources depends on aligning efforts among diverse actors and across multiple levels of governance.

1.8.3 Normative Shifts and the Politics of Cooperation

The rise of global environmental movements—including climate diplomacy—helped drive a broader normative shift toward shared responsibility, sustainability, and interdependence (Sachs, 1993). Gaining momentum from the 1970s onward, these movements reframed environmental issues as matters of justice, equity, and human security, aligning them with peacebuilding priorities (Adger et al., 2014).

Climate diplomacy, especially after the 1992 UNFCCC, stressed the need for cooperative responses to shared vulnerabilities, reinforcing the idea that environmental challenges can spur collaboration (Busby, 2022; Gleditsch, 2012).

These normative shifts reinforced the legitimacy of environmental peacebuilding by reframing natural-resource governance as a matter of shared responsibility rather than purely technical management. They likewise emphasized the importance of participatory and inclusive approaches designed to address entrenched inequalities and uneven power relations (Pretty & Ward, 2001).

1.8.4 Global Governance and the Formalization of the Field

The institutionalization of environmental peacebuilding within the United Nations and other international organizations provided the field with structure and continuity. Environmental assessments, peacekeeping guidelines, and post-conflict recovery frameworks—developed primarily in the late 1990s and 2000s—formalized the role of environmental considerations in peace and security (UNEP, 2009). These institutional milestones helped transform environmental peacebuilding from a set of isolated practices into a recognized component of international peacebuilding efforts (Bruch et al., 2016).

Institutionalization also facilitated learning and adaptation. As organizations implemented environmental interventions, they generated insights that informed subsequent policies and practices (UNDP, 2013). This iterative process contributed to the ongoing refinement of environmental peacebuilding and enhanced its effectiveness.

1.8.5 Defining the Field's Core Identity

Through the coalescence of these historical threads, environmental peacebuilding developed a distinct identity characterized by several core principles—an identity that became increasingly visible from the 2010s onward as the field matured:

- *Environmental factors mediate conflict and cooperation*. Environmental stress can heighten tensions, while shared ecological challenges can create incentives for dialogue and collaboration (Homer-Dixon, 1999; Ide, 2019).
- *Governance determines outcomes*. Institutions, rules, and decision-making processes mediate whether environmental pressures lead to conflict or cooperation (Agrawal, 2001; Ostrom, 1990).
- *Environmental interventions can support peacebuilding*. Restoration, resource governance, and sustainable livelihoods can reduce grievances, rebuild trust, and enhance stability (Bruch et al., 2016; Serrat, 2017a; Unruh, 2003).

- *Local agency is essential.* Community knowledge, participation, and ownership are critical for legitimacy, effectiveness, and long-term sustainability (Posey, 1999; Sheridan & Nyamweru, 2008).
- *Interdisciplinary frameworks are required.* Understanding environment–peace linkages demands insights from anthropology, economics, environmental science, history, international relations, political science, and related fields (Dresse et al., 2019; Matthew et al., 2002; Serrat, 2025d, 2025e, 2025f).

These principles distinguish environmental peacebuilding from related fields and provide a coherent framework for both analysis and practice.[2] By the mid-2010s, these principles were widely recognized across academic, policy, and practitioner communities. Reflecting this consolidation, global search interest in "environmental peacebuilding" has risen steadily since that period, as shown by Google Trends (https://trends.google.com/) data, indicating growing visibility and engagement.[3]

1.8.6 Looking Ahead: The Next Phase of Environmental Peacebuilding

While environmental peacebuilding has achieved significant conceptual and institutional consolidation, it continues to evolve in response to emerging challenges. Biodiversity loss, climate change, and shifting geopolitical dynamics—trends that accelerated markedly in the 2010s and 2020s—present new risks and opportunities for environmental cooperation (Adger et al., 2014; Busby, 2022). Advances in data analysis, governance models, and technology offer new tools for addressing environmental stress and supporting peacebuilding.

As policymakers, practitioners, and scholars navigate increasingly complex environmental and geopolitical landscapes, assessing environmental peacebuilding's impacts becomes essential. Since the late 2010s, evaluation frameworks have struggled to capture indirect, long-term, or relational forms of peace, as well as the nonlinear pathways through which cooperative natural-resource stewardship influences conflict dynamics (Ide, 2019). This challenge is compounded by the dominance of evaluation tools—most notably the Organisation for Economic Co-operation and Development's Development Assistance Committee's

[2] These five principles do not originate from a single source. They synthesize themes that recur across the environmental peacebuilding literature, including work on community-based resource management, environmental security, interdisciplinary approaches to environment–peace linkages, natural-resource governance, and post-conflict environmental assessment (e.g., Agrawal, 2001; Bruch et al., 2016; Dresse et al., 2019; Homer-Dixon, 1999; Posey, 1999).

[3] Google Trends data for the topic "environmental peacebuilding" (worldwide) are available from 2004. Search activity remained negligible through 2015, then rose gradually from 2016 and sharply after 2019. To date, the peak occurred in November 2025, when the topic reached a normalized value of 100. Although Google Trends reports relative rather than absolute search volume, the upward trajectory signals growing visibility and engagement with the concept.

(OECD–DAC) evaluation criteria (OECD, 2019)—that were originally designed for linear development projects rather than for complex, adaptive, and politically embedded peacebuilding processes.[4] These evaluation criteria privilege short-term, attributable outcomes and tend to overlook emergent effects, interactional change, or the multi-scalar ecological dynamics that are the hallmark of climate- and nature-based peacebuilding.

Upgrading the assessment architecture—by integrating context-sensitive indicators, mixed methods, and participatory approaches—will be critical for ensuring that environmental peacebuilding initiatives are not only well-intentioned but demonstrably impactful. Doing so requires moving beyond conventional development-oriented metrics toward evaluative frameworks capable of tracing contribution rather than attribution, capturing shifts in trust and governance, and accommodating multiple, evolving theories of change. Robust evaluation will ultimately determine the field's credibility, guide future investments, and help translate promising ideas into durable peace (Dresse et al., 2019; Serrat, 2017b). Recent work has emphasized the need for clearer principles and evaluative criteria to guide proactive environmental peacebuilding interventions (Serrat, 2025g). Emerging scholarship underscores how climate-related interventions can be intentionally designed to support peacebuilding outcomes (Serrat, 2025h). Developing assessment models that reflect this multi-layered complexity will be essential for the field's next phase.

1.9 Conclusion: From Roots to Horizons

Environmental peacebuilding did not originate from a single moment of innovation but from the gradual convergence of historical currents. Early practices of resource sharing showed that ecological interdependence could foster joint action long before the field had a name (Agrawal, 2001; Ostrom, 1990). Global environmental awareness in the late twentieth century, the rise of environmental security debates, and the expansion of sustainable development agendas broadened the conceptual and political space in which environmental issues could be linked to peace and stability (Barnett, 2001; Brauch, 2008). In parallel, climate-focused diplomacy and transnational environmental activism increasingly cast ecological challenges as shared threats, reinforcing the view that effective environmental governance is fundamentally intertwined with broader issues of human well-being and security (Biermann & Pattberg, 2008; Sachs, 1993).

Institutional developments—particularly within the United Nations system—played a decisive role in transforming these ideas into practice. UNEP's post-conflict environmental assessments, the integration of natural resource governance into peacekeeping and recovery mandates, and the evolution of the peacebuilding

[4]The OECD–DAC framework encompasses relevance, coherence, effectiveness, efficiency, impact, and sustainability.

architecture of the United Nations helped formalize the role of environmental considerations in conflict-affected settings (Bruch et al., 2016; UNDP, 2013; UNEP, 2009). These efforts provided the first systematic evidence base demonstrating that environmental management could support stabilization, recovery, and long-term peace.

Concurrently, developments in academic research deepened the field's analytical underpinnings. Interdisciplinary scholarship, methodological innovation, and critical debate clarified the mechanisms linking environmental risk factors, cooperation, and conflict (Homer-Dixon, 1999; Ide, 2019; Kahl, 2006). Research networks that grew throughout the late 1990s and 2000s were instrumental in translating empirical insights into practical policy guidance, grounding environmental peacebuilding approaches in the lived realities and needs of affected communities (Conca & Dabelko, 2002; Matthew et al., 2002). By the 2010s, these intellectual and institutional developments had consolidated into a recognizable area of inquiry with its own principles, practices, and professional community (Dresse et al., 2019). As the field matured, the need for systematic evaluation became increasingly apparent, both to demonstrate impact and to guide more effective interventions.

As environmental peacebuilding continues to evolve, its future trajectory will be shaped by accelerating ecological change, shifting geopolitical dynamics, and rising expectations for interventions that are both effective and equitable (Adger et al., 2014; Busby, 2022; Ide et al., 2021). Strengthening evaluation frameworks—especially those capable of capturing long-term, relational, and context-specific dimensions of peace—will be crucial for demonstrating the field's impact (Ide, 2019). Equally vital is expanding the integration of local knowledge, participatory governance, and interdisciplinary inquiry so that environmental peacebuilding stays context-attuned, inclusive, and methodologically rigorous (Posey, 1999; Pretty & Ward, 2001).

In this sense, the future of environmental peacebuilding will depend on the same forces that shaped its past: recognizing that environmental governance is inseparable from human security, sustaining collaboration across actors and scales, and maintaining the capacity for adaptation (Adger et al., 2014). Yet the accelerating pace of ecological disruption—from rapid biodiversity loss to intensifying climate impacts—means that these commitments can no longer be treated as aspirational ideals but as urgent imperatives. A historical perspective underscores that environmental peacebuilding has advanced most effectively when ecological stressors, political will, and governance innovation align. Over the coming decade, the top three concerns identified in the World Economic Forum's 2026 Global Risks Perception Survey all relate to environmental threats—extreme weather events, biodiversity loss, and ecosystem collapse—all relate to environmental threats, bringing into sharp relief how ecological disruption has become the defining long-term challenge to global stability (World Economic Forum, 2026). As environmental risks intensify and geopolitical tensions sharpen, the need for effective, equitable, and empirically grounded environmental peacebuilding becomes increasingly pressing. Whether the field can meet this moment will depend on its ability to draw on its deep historical lineage while adapting to unprecedented environmental and geopolitical realities. The choices made in the coming years will determine whether environmental pathways to cooperation can keep pace with mounting pressures, or whether widening gaps between environmental change and governance responses undermine the sustainability of shared ecosystems—and, with them, the prospects for lasting peace and stability.

References

Adger, W., Pulhin, J., & Barnett, J. (2014). Human security. In *IPCC Fifth Assessment Report* (pp. 755–791). Cambridge University Press.

Agrawal, A. (2001). Common property institutions and sustainable governance of resources. *World Development, 29*(10), 1649–1672.

Ali, S. (Ed.). (2007). *Peace parks: Conservation and conflict resolution*. MIT Press.

Autesserre, S. (2014). *Peaceland: Conflict resolution and the everyday politics of international intervention*. Cambridge University Press.

Bankoff, G. (2003). *Cultures of disaster: Society and natural hazard in the Philippines*. Routledge.

Barnett, J. (2001). *The meaning of environmental security: Ecological politics and policy in the new security era*. Zed Books.

Behnke, R., Scoones, I., & Kerven, C. (Eds.). (1993). *Range ecology at disequilibrium: New models of natural variability and pastoral adaptation in African savannas*. ODI/IIED.

Berkes, F. (1985). Fishermen and the "tragedy of the commons". *Environmental Conservation, 12*(3), 199–206.

Berkes, F. (1989). *Common property resources: Ecology and community-based sustainable development*. Belhaven Press.

Biermann, F., & Pattberg, P. (2008). Global environmental governance: Taking stock, moving forward. *Annual Review of Environment and Resources, 33*, 277–294.

Brauch, H. (2008). Environment and human security: Towards freedom from hazard impacts. *InterSecTions, 2*, 1–55.

Bruch, C., Muffett, C., & Nichols, S. (Eds.). (2016). *Governance, natural resources, and post-conflict peacebuilding*. Routledge.

Busby, J. (2022). *States and nature: The effects of climate change on security*. Cambridge University Press.

Conca, K., & Dabelko, G. (Eds.). (2002). *Environmental peacemaking*. Johns Hopkins University Press.

Dinar, S., Dinar, A., McCaffrey, S., & McKinney, D. (2007). *Bridges over water: Understanding transboundary water conflict, negotiation and cooperation*. World Scientific.

Dresse, A., Nielsen, J., Zikos, D., & Cormier, R. (2019). Environmental peacebuilding: Towards a theoretical framework. *Cooperation and Conflict, 54*(1), 99–119.

Gleditsch, N. (2012). Whither the weather? Climate change and conflict. *Journal of Peace Research, 49*(1), 3–9.

Homer-Dixon, T. (1999). *Environment, scarcity, and violence*. Princeton University Press.

Ide, T. (2019). The impact of environmental cooperation on peacemaking: Definitions, mechanisms, and empirical evidence. *International Studies Review, 21*(3), 327–346.

Ide, T., Bruch, C., Carius, A., Conca, K., Dabelko, G., Matthew, R., & Weinthal, E. (2021). The past and future(s) of environmental peacebuilding. *International Affairs, 97*(1), 1–16.

Kahl, C. (2006). *States, scarcity, and civil strife in the developing world*. Princeton University Press.

Le Billon, P. (2005). *Fuelling war: Natural resources and armed conflicts*. Routledge.

Lederach, J. (1997). *Building peace: Sustainable reconciliation in divided societies*. United States Institute of Peace Press.

Mach, K., Kraan, C., Adger, W., Buhaug, H., Burke, M., Fearon, J., et al. (2019). Climate as a risk factor for armed conflict. *Nature, 571*, 193–197.

Matthew, R., Halle, M., & Switzer, J. (Eds.). (2002). *Conserving the peace: Resources, livelihoods and security*. International Institute for Sustainable Development.

Mitchell, R. (2003). International environmental agreements: A survey of their features, formation, and effects. *Annual Review of Environment and Resources, 28*, 429–461.

Mithen, S. (2012). *Thirst: Water and power in the ancient world*. Harvard University Press.

Netting, R. M. (1993). *Smallholders, householders: Farm families and the ecology of intensive, sustainable agriculture*. Stanford University Press.

Niamir-Fuller, M. (Ed.). (1999). *Managing mobility in African rangelands: The legitimization of transhumance*. Intermediate Technology Publications.

OECD. (2019). *Better criteria for better evaluation: Revised evaluation criteria definitions and principles for use*. OECD Publishing. https://www.oecd.org/en/publications/2019/12/better-criteria-for-better-evaluation_f7a307eb.html

Ostrom, E. (1990). *Governing the commons: The evolution of institutions for collective action*. Cambridge University Press.

Paris, R. (2004). *At war's end: Building peace after civil conflict*. Cambridge University Press.

Posey, D. (1999). *Cultural and spiritual values of biodiversity*. UNEP.

Pretty, J., & Ward, H. (2001). Social capital and the environment. *World Development, 29*(2), 209–227.

Rustad, S., & Binningsbø, H. (2012). A price worth fighting for? Natural resources and conflict recurrence. *Journal of Peace Research, 49*(4), 531–546.

Sachs, W. (Ed.). (1993). *Global ecology: A new arena of political conflict*. Zed Books.

Serrat, O. (2017a). The sustainable livelihoods approach. In *Knowledge solutions: Tools, methods, and approaches to drive organizational performance* (pp. 21–26). Springer.

Serrat, O. (2017b). Learning from evaluation. In *Knowledge solutions: Tools, methods, and approaches to drive organizational performance* (pp. 879–884). Springer.

Serrat, O. (2025a). ICYMI: Climate security rhymes with national security. In *Anthropogenic solutions for climate change: Achieving environmental peace* (pp. 31–34). Springer.

Serrat, O. (2025b). Preserving paradise: The Heart of Borneo Initiative. In *Anthropogenic solutions for climate change: Achieving environmental peace* (pp. 221–227). Springer.

Serrat, O. (2025c). Safeguarding the mother of waters: Environmental peacebuilding in the Mekong River Basin. In *Anthropogenic solutions for climate change: Achieving environmental peace* (pp. 229–239). Springer.

Serrat, O. (2025d). Disciplinary perspectives on environmental peacebuilding: Insights from environmental science and international relations. In *Anthropogenic solutions for climate change: Achieving environmental peace* (pp. 147–164). Springer.

Serrat, O. (2025e). *Bridging the gap: Synergizing anthropology, economics, history, and sociology for environmental peacebuilding*. Unpublished manuscript, Georgetown University. https://www.researchgate.net/publication/391522222_Bridging_the_Gap_Synergizing_Anthropology_Economics_History_and_Sociology_for_Environmental_Peacebuilding

Serrat, O. (2025f). *Accounting for complexity: Liberal studies and environmental peacebuilding*. Unpublished manuscript, Georgetown University. https://www.researchgate.net/publication/396824279_Accounting_for_Complexity_Liberal_Studies_and_Environmental_Peacebuilding

Serrat, O. (2025g). *Assessing environmental peacebuilding: Principles, criteria, and pathways for proactive interventions*. Unpublished manuscript, Georgetown University. https://www.researchgate.net/publication/398543700_Assessing_Environmental_Peacebuilding_Principles_Criteria_and_Pathways_for_Proactive_Interventions

Serrat, O. (2025h). *Anthropogenic solutions for climate change: Achieving environmental peace*. Springer.

Sheridan, M., & Nyamweru, C. (Eds.). (2008). *African sacred groves: Ecological dynamics and social change*. James Currey.

UNDP. (2013). *Natural resource management in transition settings*. United Nations Development Programme.
UNEP. (2009). *From conflict to peacebuilding: The role of natural resources and the environment*. United Nations Environment Programme.
Unruh, J. (2003). Land tenure and legal pluralism in postwar societies. *Peace & Change, 28*(3), 352–377.
Unruh, J., & Williams, R. (Eds.). (2013). *Land and post-conflict peacebuilding*. Routledge.
Wisner, B., Blaikie, P., Cannon, T., & Davis, I. (2004). *At risk: Natural hazards, people's vulnerability and disasters* (2nd ed.). Routledge.
World Economic Forum. (2026). *Global Risks Report 2026*. World Economic Forum.
Wolf, A. (2007). Shared waters: Conflict and cooperation. *Annual Review of Environment and Resources, 32*, 241–269.
Zeitoun, M., & Mirumachi, N. (2008). Transboundary water interaction I: Reconsidering conflict and cooperation. *International Environmental Agreements, 8*(4), 297–316.

Chapter 2
Disciplinary Perspectives on Environmental Peacebuilding: Insights from Environmental Science and International Relations

Abstract By integrating environmental considerations into peacebuilding initiatives, we can create lasting solutions that address recurring root causes of conflict, such as competition over natural resources, environmental degradation, and resource scarcity. Toward this, we must adopt a holistic and cross-disciplinary approach, embracing diverse viewpoints and integrating multiple areas of expertise. This précis delves into how insights from environmental science and international relations might specifically contribute to environmental peacebuilding.

2.1 Harnessing Nature for Peace: The Power of Environmental Peacebuilding

Envision a world where nature's elements—air, minerals, soil, water, and other resources—serve as the keystones of peace (Conca & Dabelko, 2002). This transformative approach, environmental peacebuilding, addresses shared environmental challenges to promote cooperation, trust, and sustainable development (United Nations Environment Programme, 2023).

By integrating environmental considerations into peacebuilding initiatives, we can create lasting solutions that target the root causes of conflict, such as competition over natural resources, environmental degradation, and resource scarcity (Homer-Dixon, 1999). Examples of environmental peacebuilding initiatives include: climate change adaptation projects, collaborative water management, community-based natural resource management, cross-border wildlife conservation, forest conservation and management, joint environmental monitoring, marine conservation agreements, reforestation and land restoration projects, renewable energy projects, sustainable agriculture programs, transboundary air pollution agreements, and

Previously published in Serrat, O. (2025). *Anthropogenic solutions for climate change: Achieving environmental peace*. Springer.

O. Serrat, *Advancing Environmental Peacebuilding*, SpringerBriefs in Political Science, https://doi.org/10.1007/978-981-92-4028-9_2

transboundary protected areas. This holistic approach addresses the intertwined challenges of environmental sustainability and social stability, ensuring that communities can thrive with sustainable livelihoods and a healthy environment (Serrat, 2017). By fostering cooperation, trust, and resilience, we can lay the foundation for enduring peace and long-term stability (Dresse et al., 2019; Heffington et al., 2019a; Ide, 2021; Krampe et al., 2024; Scoones, 1998; United Nations Environment Programme, 2023).

Unavoidably, effective environmental peacebuilding initiatives require inclusive stakeholder engagement, robust collaboration, innovative strategies, continuous adaptation, and thorough evaluation of lessons learned (Scoones, 1998; Serrat, 2025a). Each initiative should thoroughly assess the strategic situation, define the desired outcomes, develop the necessary means to achieve these outcomes, design ways to leverage these means effectively, and assess the associated costs and risks (Heffington et al., 2019b).

Above all, to achieve success, environmental peacebuilding must adopt a holistic and cross-disciplinary approach, embracing diverse viewpoints and integrating multiple areas of expertise (Sommer & Fassbender, 2024). While specialized knowledge is valuable, relying solely on the reductionism that single disciplines encourage can oversimplify complex problems by analyzing them in isolation. Anchored in positivism, pragmatism, and realism, and employing reductionism when necessary, the deepest insights often emerge through abduction, deduction, and induction at the crossroads of various disciplines, with pragmatism often proving particularly effective (Slingerland, 2008). In environmental peacebuilding, as in other areas in general, single disciplines should complement each other rather than compete for supremacy. Environmental peacebuilding is inherently integrative, drawing insights from both the natural and social sciences to understand and manage the intersections of environmental challenges and conflict dynamics, and to create solutions that are both sustainable and transformative. For instance, enhanced academic rigor suggests the combined efforts of environmental science and international relations, as examples discussed in this précis, can significantly upgrade the effectiveness of environmental peacebuilding.

2.2 Research Question

According to the Global Atlas of Environmental Justice (https://ejatlas.org/), there were more than 4200 documented environmental conflicts worldwide as of early 2024.[1] Given the global significance of environmental conflicts and the critical roles

[1] The Global Atlas of Environmental Justice (https://ejatlas.org/) classifies environmental conflicts across 10 categories: biodiversity conservation conflicts, biomass and land conflicts (forests, agriculture, fisheries and livestock management), fossil fuels and climate justice/energy, infrastructure and built environment, industrial and utilities conflicts, mineral ores and building materials extrac-

that both environmental science and international relations can play in understanding and addressing these issues, this précis delves into how insights from these two disciplines might specifically contribute to environmental peacebuilding.[2] Environmental science provides crucial knowledge about natural systems and sustainable management practices, while international relations offer insights into the economic, political, and social dynamics of peace and conflict.[3] By combining these perspectives, we can develop more comprehensive and effective strategies for environmental peacebuilding.

This précis delves into the question: How can environmental science and international relations contribute to environmental peacebuilding initiatives? It explores how these disciplines address issues from different perspectives, uniquely illuminate the research question, provide answers, and diverge in their investigative approaches. Finally, it suggests ways to integrate these perspectives to make progress on the research question.

---

tion, nuclear, tourism recreation, waste management, and water management. The Global Atlas of Environmental Justice (https://ejatlas.org/) is a work in progress: newly documented cases and information are continuously added. However, many conflicts remain undocumented and new ones arise.

[2] Environmental peacebuilding initiatives are not solely international. While many conflicts involve shared resources that cross national borders—such as air quality, forests, groundwater aquifers, marine ecosystems, mountain ranges, rivers and lakes, and wildlife migration corridors—peacebuilding can also occur within a single country. Local factors driving peacebuilding include climate change, cultural and historical significance, economic development, policy effectiveness, environmental degradation, religious and ethnic tensions, resource scarcity, and social tensions (Peace Direct & Alliance for Peacebuilding, 2020). This précis focuses on transboundary peacebuilding initiatives to examine the specific role of international relations.

[3] International relations, a subfield of political science, addresses issues such as conflict, diplomacy, global economic relations, global governance, human rights, international law, and war (Mingst & Arreguín-Toft, 2016). Political science, the broader field, explores the theory and practice of politics and government at all levels—local, national, and international—and includes other subfields such as comparative politics, political theory, public administration, and public policy (Roskin & Cord, 2019). This précis highlights the relevance of international relations to environmental peacebuilding due to its emphasis on interactions between countries and global challenges that transcend national borders (Goldstein & Pevehouse, 2017). These challenges encompass environmental issues, conflict and cooperation, international agreements and treaties, the role of international organizations, and diplomacy and foreign policy. In summary, political science provides the broader context of governance and policy, while international relations offers tools and frameworks to address complex, transboundary environmental challenges and the collaborative efforts required for peacebuilding.

2.3 Quid Environmental Peacebuilding?

2.3.1 *Emergence and Evolution of Environmental Peacebuilding*

Environmental peacebuilding emerged in the 1980s, recognizing that competition over natural resources, environmental degradation, and resource scarcity can exacerbate social tensions and contribute to conflict (Homer-Dixon, 1999). This concept is particularly relevant in today's interconnected world, where climate change, deforestation, water scarcity, and other environmental challenges threaten local and global stability (Heffington et al., 2019b). For example, overfishing in coastal regions disrupts livelihoods, leading to economic hardship and social unrest. Deforestation in the Amazon affects local social unrest, biodiversity, and global climate change, impacting ecological balance. Climate change impacts are global, causing displacement, food insecurity, and heightened tensions. These examples underscore the urgent need for environmental peacebuilding as a comprehensive approach to addressing the multifaceted challenges of environmental conflicts at both local and global levels (Dresse et al., 2018).

2.3.2 *The Complex Interplay Between Environmental Factors and Conflict Dynamics*

Initially, environmental peacebuilding focused on understanding the impact of environmental factors on conflict dynamics. The field has since evolved to address the complex interplay between environmental factors and conflict dynamics, incorporating strategies for leveraging environmental management to promote peace (Krampe et al., 2024; Krampe & Swain, 2024).

To address the multifaceted challenges of environmental conflicts, environmental peacebuilding must integrate insights from multiple disciplines, requiring an understanding of cultural, ecological, economic, political, and social dimensions.[4]

[4] For example, initiatives aimed at reducing deforestation in the Amazon often face resistance due to conflicting conservation and economic interests (e.g., logging) that bypass legislation. Collaborative efforts involving indigenous leaders, ecologists, economists, and sociologists using real-time satellite monitoring and stricter penalties for violations could create sustainable strategies that protect ecosystems. In Colombia, post-conflict efforts to restore degraded lands and integrate former combatants into conservation work face logistical and social challenges. Environmental scientists, social workers, and conflict resolution specialists could energize these initiatives and boost their long-term success. In the Lake Chad region, environmental degradation and water scarcity have worsened conflicts among communities reliant on the lake, which spans four countries (Serrat, 2025b). A cross-disciplinary approach involving diplomats, climate scientists, agricultural experts, and conflict resolution specialists could address both environmental and social aspects, offering a path to resilience. In the Mekong River basin, dam construction has fueled

Fig. 2.1 Principal Disciplines Relevant to Environmental Peacebuilding. Note. This figure illustrates the principal disciplines contributing to environmental peacebuilding. Other relevant disciplines may include environmental education, environmental engineering, environmental ethics, environmental psychology, geology, hydrology, public health, sociology, and urban planning, depending on the context

This cross-disciplinary approach fosters a comprehensive understanding of environmental factors and conflict dynamics. By harnessing the strengths of multiple disciplines, environmental peacebuilding can create a more robust framework for addressing the environmental challenges that contribute to conflict while fostering lasting peace and stability. Figure 2.1 identifies the core disciplines that contribute to environmental peacebuilding, emphasizing the interconnected nature of ecological sustainability and conflict resolution.

2.3.3 *Core Goals of Environmental Peacebuilding*

Necessarily, environmental peacebuilding initiatives must be tailored to address specific local environmental, political, and social contexts. Therefore, these initiatives aim to tackle diverse challenges through various goals (Krampe & Swain, 2024). Drawing on Matthew et al. (2009), Conca and Dabelko (2002), and Conca and Wallace (2009), it is possible to identify core goals that environmental peacebuilding initiatives typically seek to advance:

---

international disputes over water rights, ecological damage, and impacts on fisheries and agriculture (Serrat, 2025c). Cross-disciplinary teams of hydrologists, environmental scientists, diplomats, and local representatives could strengthen transboundary water governance. In every scenario, stronger political will could turn challenges into opportunities.

- **Trust and Cooperation.** Collaborative environmental projects foster trust and cooperation between adversaries, laying the foundation for peace and stability.
- **Conflict Prevention and Resolution.** By managing shared natural resources and addressing environmental issues collaboratively, communities and nations can reduce tensions and prevent conflicts arising from competition over natural resources, environmental degradation, and resource scarcity.
- **Environmental Justice.** Environmental justice ensures fair and equitable treatment of all individuals, including future generations, by considering their needs and rights in the development, implementation, and enforcement of environmental policies and practices.
- **Inclusive Governance.** Environmental peacebuilding involves inclusive decision-making that engages diverse stakeholders, including marginalized groups, ensuring their voices are heard and needs addressed.
- **Resilience and Adaptation.** Strengthening environmental resilience and promoting adaptive strategies help communities cope with environmental challenges, reducing future conflict likelihood.
- **Sustainable Development.** Environmental peacebuilding promotes sustainable development by managing natural resources to benefit both the environment and local communities. Figure 2.2 lists the core goals that environmental peacebuilding initiatives typically seek to advance.

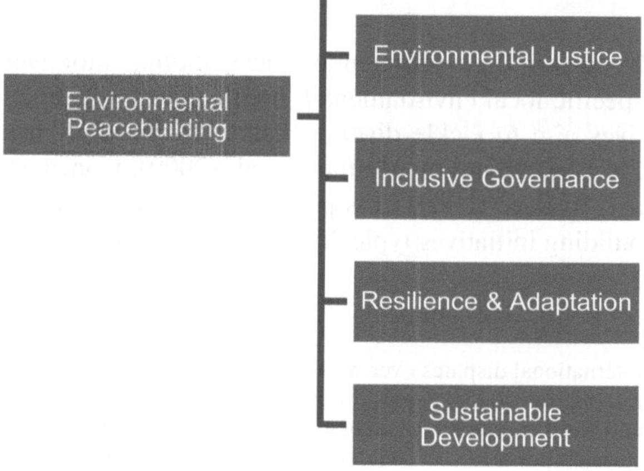

Fig. 2.2 Core Goals of Environmental Peacebuilding Initiatives. Note. This figure lists core goals guiding environmental peacebuilding initiatives. These goals aim to address various challenges and promote peace and stability through inclusive and sustainable approaches. They are not necessarily mutually inclusive and can independently contribute to the success of an initiative

2.4 Theoretical Foundations of Environmental Science and International Relations

To explore how environmental science and international relations contribute to environmental peacebuilding, we must examine their theoretical foundations (Bennett & Chapin, 2018; Sommer & Fassbender, 2024). This understanding of key concepts, principles, and perspectives enables us to identify effective strategies for fostering environmental peace and stability. By delving into their theoretical underpinnings, we can uncover synergies and complementarities to address complex environmental challenges and promote sustainable, long-lasting peace.

2.4.1 Key Principles of Environmental Science

Environmental science is grounded in the study of natural systems and the interactions between organisms and their environment (Bennett & Chapin, 2018; Odum, 1971). It rests on principles from ecology, sustainability, and systems thinking to address complex environmental challenges. Drawing on Withgott and Laposata (2012), some key principles informing environmental science are:

- **Climate Science:** The study of climate systems, including the causes and effects of climate change, and the development of strategies to mitigate and adapt to its impacts (National Aeronautics and Space Administration, n.d.).
- **Ecology:** The study of interactions between organisms and their environment, focusing on ecosystems, biodiversity, and the flow of energy and nutrients (Begon et al., 2006).
- **Environmental Ethics:** The philosophical study of moral values and principles concerning the environment, including the rights of non-human entities and human responsibilities towards nature (Callicott & Frodeman, 2009).
- **Sustainability:** Meeting present needs without compromising future generations, framed around economic, environmental, and social sustainability (United Nations, n.d.).
- **Systems Thinking:** Viewing problems as part of an overall system rather than isolated incidents. It emphasizes the interconnectedness of components and considers how changes in one part can affect the whole (Meadows, 2008).

2.4.2 Key Perspectives in International Relations

International relations examines the dynamics of global politics, especially the interplay between nations, non-state actors, and international institutions (Kydd, 2015; Mingst & Arreguín-Toft, 2016). Informed by diverse theoretical perspectives like constructivism, liberalism, and realism, the field analyzes and addresses issues of cooperation, conflict, and power globally. Drawing on Reus-Smit & Snidal (2008), some key perspectives informing international relations are:

- **Constructivism:** The role of ideas, norms, and identities in shaping international relations, arguing that state behavior is influenced by social constructs and shared beliefs (Adler, 2019).
- **Feminism:** Examines how gender shapes international relations, advocating for the inclusion of women's experiences and voices in global politics (True, 2018).
- **Liberalism:** Emphasizes cooperation, democracy, and international institutions as means to achieve peace and stability, advocating for economic interdependence and multilateralism (Ikenberry, 2018).
- **Marxism:** Analyzes international relations through the lens of economic inequality and class struggle, highlighting the impact of capitalism on global politics (Rupert & Smith, 2016).
- **Realism:** Views international relations as a struggle for power among self-interested states, emphasizing national security and military strength (Mearsheimer, 2018).

These theoretical foundations provide a framework for understanding the complex interactions between environmental issues and international relations, highlighting the importance of cross-disciplinary approaches to address global challenges.

2.5 Disciplinary Insights from Environmental Science and International Relations

2.5.1 Environmental Science: Key Concepts and Contributions

Environmental scientists have long studied how human activities impact ecosystems and how environmental degradation can lead to social unrest. From deforestation in the Amazon to water shortages in the Middle East, the link between environmental stress and conflict is well-documented (Bennett & Chapin, 2018). Environmental science provides insights into peacebuilding's ecological aspects, such as conservation strategies, natural resource management, and the role of biodiversity in promoting stability and resilience. By revealing how environmental changes can trigger or escalate conflicts, it offers strategies to mitigate these risks through sustainable management and restoration efforts (Sommer & Fassbender, 2024).

Environmental science uniquely contributes to the research question by providing crucial insights into the natural systems and processes underpinning environmental peacebuilding (Sommer & Fassbender, 2024). Drawing on Cunningham and Cunningham (2022), here are key ways in which environmental science can inform environmental peacebuilding initiatives:

- **Climate Science.** Insights from climate science identify the impacts of climate change on conflict regions. Understanding challenges like increased droughts or rising sea levels allows scientists to recommend adaptive measures to mitigate impacts and reduce conflict risk.
- **Ecological Understanding.** Environmental science offers a deep understanding of ecosystems, biodiversity, and organism-environment interactions. This knowledge is essential for developing strategies to manage resources sustainably and restore conflict-degraded ecosystems.
- **Environmental Ethics.** Environmental science often incorporates environmental ethics, guiding decision-making regarding non-human entities' rights and humans' responsibilities towards nature. This ethical framework supports inclusive and equitable environmental peacebuilding initiatives.
- **Environmental Monitoring.** Advanced monitoring techniques allow scientists to track environmental changes in real-time. This data assesses ecosystem health, detects early signs of stress, and implements timely interventions to prevent conflict.
- **Restoration Ecology.** Environmental science includes restoration ecology, focusing on rehabilitating degraded environments. Restoring ecosystems can rebuild livelihoods, promote social stability, and provide a foundation for long-term peace in post-conflict regions.
- **Sustainable Resource Management.** Environmental science provides methodologies for sustainably managing resources like water, forests, and minerals. Effective resource management prevents resource scarcity, often a conflict driver, and promotes equitable access, fostering community cooperation.

2.5.2 International Relations: Key Concepts and Contributions

International relations has traditionally focused on state interactions but increasingly recognizes the importance of non-state actors and transnational issues like the environment (Kydd, 2015). Researchers examine how environmental issues influence conflict, international cooperation, and state behavior (Barnett, 2001). The concept of environmental security highlights how environmental challenges can undermine national security and stability. International relations provides a framework for understanding the role of diplomacy, international organizations, and treaties in addressing environmental conflicts and promoting peace.

International relations uniquely contributes to answering the research question by providing insights into the economic, political, and social dynamics of environmental peacebuilding (Sommer & Fassbender, 2024). Drawing on Bodansky et al. (2007), here are key ways in which international relations can inform environmental peacebuilding initiatives:

- **Conflict Prevention and Resolution.** International relations provides frameworks for conflict prevention and resolution, focusing on diplomacy, mediation, and negotiation. This helps design strategies to de-escalate tensions and promote peaceful coexistence.

- **Diplomatic Negotiations.** International relations helps understand how diplomatic negotiations and treaties manage shared environmental resources and address conflicts. For example, international agreements on water rights can help prevent disputes over water scarcity.
- **Economic Interdependence.** International relations appreciates the role of economic interdependence in promoting peace and stability. By fostering trade and economic cooperation, international relations help create conditions conducive to environmental peacebuilding.
- **Global Governance.** International relations explores the structures and mechanisms of global governance, including international environmental regimes and conventions. These mechanisms establish norms, standards, and regulations guiding environmental management and peacebuilding efforts.
- **Human Security.** International relations emphasizes the concept of human security, which includes environmental security as a key component. This perspective underscores the importance of protecting individuals and communities from environmental threats and ensuring their well-being.
- **International Cooperation.** International relations highlights the importance of international cooperation in addressing global environmental challenges. This includes the role of international organizations, such as the United Nations, in facilitating collaboration and coordination to mitigate environmental conflicts.
- **Power Dynamics.** International relations examines power dynamics and the influence of various state and non-state actors on environmental policies. Understanding these dynamics is crucial for developing effective environmental peacebuilding initiatives that account for different stakeholders' interests and actions.

2.5.3 Comparative Analysis of Environmental Science and International Relations

Naturally, given their distinct disciplinary origins, environmental science and international relations diverge significantly in their investigative approaches (Barnett, 2018).

2.5.3.1 The Practice of Environmental Science

Environmental science is driven by cross-disciplinary collaboration, empirical and data-driven approaches, a focus on natural systems, and quantitative analysis to address environmental challenges (Cunningham & Cunningham, 2022):

- **Cross-Disciplinary Collaboration.** Environmental science integrates knowledge from biology, chemistry, physics, and geology to understand complex environmental phenomena. This approach is essential for addressing multifaceted environmental issues.
- **Empirical and Data-Driven.** Environmental science relies on empirical data and scientific methods to study natural systems. Researchers conduct field studies, experiments, and use advanced technologies like remote sensing and geographic information systems to collect and analyze data.
- **Focus on Natural Systems.** Environmental science aims to understand interactions within natural systems, including ecosystems, climate systems, and the impact of human activities. The goal is to develop solutions for conservation, sustainability, and restoration.
- **Quantitative Analysis.** Environmental science employs quantitative analysis to measure environmental variables, model ecological processes, and predict future scenarios. This includes statistical analysis, mathematical modeling, and simulation.

2.5.3.2 The Practice of International Relations

International relations is driven by case studies and comparative analysis, focusing on governance and diplomacy, understanding political and social contexts, and using qualitative and theoretical analysis (Dunne et al., 2021; Mingst & Arreguín-Toft, 2016).

- **Case Studies and Comparative Analysis.** International relations frequently uses case studies and comparative analysis to examine instances of international cooperation, conflict, and diplomacy. This approach helps identify patterns, draw lessons, and develop theories.
- **Focus on Governance and Diplomacy.** International relations investigates international governance mechanisms, the role of diplomacy, and the impact of international policies and agreements. The goal is to develop strategies for conflict prevention, resolution, cooperation, and promoting global stability.
- **Political and Social Context.** International relations emphasizes political and social contexts in shaping international interactions. Researchers examine the influence of cultural factors, ideologies, and power dynamics on global politics.
- **Qualitative and Theoretical Analysis.** International relations often relies on qualitative analysis and theoretical frameworks to study the behavior of states, international organizations, and non-state actors. Researchers analyze historical events, refer to policy documents, and conduct interviews to understand motivations and strategies in the international arena.

2.5.3.3 Contrasting Investigative Methods: Environmental Science and International Relations

Environmental science and international relations approach the research question from different angles, reflecting their unique focal points, guiding principles, integrative approaches, methodologies, and perspectives (Barnett, 2018). Key areas of contrast include:

- **Focus Areas.** Environmental science converges on natural systems, ecological processes, and the physical environment. It aims to understand the natural world and develop solutions for environmental conservation, sustainability, and restoration. Conversely, international relations concentrates on the behavior of states, international organizations, and non-state actors in the global political arena. It seeks to understand the dynamics of power, cooperation, and conflict among these actors.
- **Guiding Principles.** Environmental science is grounded in scientific principles and emphasizes objectivity, empirical evidence, and the interconnectedness of natural systems. The goal is to develop scientifically-based solutions for environmental challenges. International relations is rooted in political theories and emphasizes the role of power dynamics, diplomacy, and governance. The goal is to develop strategies for conflict prevention, resolution, international cooperation, and global stability.
- **Integrative Approaches.** Environmental science integrates knowledge from various scientific disciplines such as biology, chemistry, and geology. International relations draws from political science, sociology, economics, and history to understand complex global interactions.
- **Research Methods.** Environmental science uses empirical data, field studies, laboratory experiments, and quantitative analysis to study environmental phenomena. Techniques like remote sensing, geographic information systems, and ecological modeling are commonly employed. International relations relies on theoretical frameworks, historical analysis, case studies, and qualitative methods. Researchers analyze policy documents, conduct interviews, and use comparative analysis to understand international interactions.
- **Viewpoints.** Environmental science views environmental challenges through the lens of ecological impact and sustainability, emphasizing preserving biodiversity and maintaining healthy ecosystems. International relations examines environmental challenges through the lens of political and economic impact, considering how environmental issues affect state behavior, international relations, and global security.

Given their differing focus areas, guiding principles, integrative approaches, research methods, and viewpoints, it is not surprising that the solutions proposed by environmental science and international relations vary significantly. In the context of environmental peacebuilding, environmental science typically offers technical and scientific solutions such as conservation strategies, ecosystem restoration, and sustainable resource management, prioritizing the environmental root causes of conflict (Millennium Ecosystem Assessment, 2005). In contrast, international relations usually suggest policy recommendations, diplomatic strategies, and frameworks for international cooperation, focusing on managing political and social dynamics to prevent and resolve conflicts (Conca & Dabelko, 2002). By understanding these divergent approaches, we can better appreciate each discipline's unique contributions to environmental peacebuilding and, where possible, combine them to create more comprehensive and effective initiatives.

2.6 Practical Implications of Integrating Disciplinary Insights

The integration of insights from environmental science and international relations in environmental peacebuilding initiatives has several practical implications (Conca & Dabelko, 2002). Key implications include:

- **Effective Community Engagement**
  - **Capacity Building.** Insights from both disciplines support capacity-building initiatives that strengthen local communities' and institutions' ability to manage environmental resources sustainably. This reduces dependency on external assistance and promotes long-term resilience.
  - **Inclusive Participation.** Environmental science highlights involving local communities in environmental management, ensuring their knowledge and needs are considered. International relations emphasize stakeholder engagement and participatory approaches in decision-making. Together, these disciplines promote inclusive participation, empowering communities, and enhancing the legitimacy of environmental peacebuilding initiatives.
- **Enhanced Understanding of Conflict Drivers.** Combining the ecological knowledge of environmental science with the political and social insights of international relations helps develop a comprehensive understanding of environmental conflicts' underlying drivers. This holistic analysis can identify the root causes and dynamics of conflicts, leading to more targeted and effective interventions.
- **Improved Policy and Strategy Development**
  - **Adaptive Management.** Environmental science contributes methodologies for monitoring and assessing environmental conditions, while international relations provide frameworks for diplomatic engagement and conflict prevention or resolution. Synergizing the two disciplines enables adaptive management approaches based on real-time data and changing political contexts.
  - **Informed Decision-Making.** Integrating environmental science and international relations allows for creating policies and strategies informed by scientific data and political realities. This ensures that peacebuilding initiatives are grounded in empirical evidence and are politically feasible, increasing their likelihood of success.

- **Strengthened International Cooperation**
  - **Regional and Global Initiatives.** Synergizing environmental science and international relations enables developing regional and global initiatives addressing transboundary environmental issues. This fosters collaboration among neighboring countries and international organizations, reducing tensions and building trust.
  - **Shared Environmental Goals.** Environmental science emphasizes sustainable resource management, serving as common ground for cooperation among conflicting parties. International relations facilitate diplomatic endeavors and international agreements supporting collaborative environmental management, promoting peace and stability.
- **Sustainable Development and Resilience**
  - **Long-Term Solutions.** Environmental science provides technical solutions for ecosystem restoration and sustainable resource management, while international relations ensure these solutions are embedded within a supportive political framework. Harnessing both disciplines leads to sustainable development outcomes addressing environmental and social dimensions of conflict.
  - **Resilience Building.** By tackling the environmental root causes of conflict and promoting cooperative political strategies, integrated peacebuilding initiatives enhance communities' resilience to environmental and socio-political shocks. This reduces the risk of conflict re-emergence and contributes to lasting peace.
- **Technological Innovations**
  - **Advanced Monitoring and Evaluation.** Utilizing technologies such as data analytics, geographic information systems, and remote sensing enhances the monitoring and evaluation of environmental peacebuilding initiatives. These technologies provide accurate and timely data, enabling adaptive management and informed decision-making.
  - **Sustainable Technologies.** Research and implementation of sustainable technologies for climate adaptation, ecosystem restoration, and resource management significantly contribute to the success of environmental peacebuilding initiatives.

In summary, the integration of environmental science and international relations in environmental peacebuilding initiatives offers a robust and multifaceted approach to addressing the complex challenges posed by environmental conflicts. This cross-disciplinary collaboration enhances the effectiveness, resilience, and sustainability of environmental peacebuilding initiatives (Conca & Dabelko, 2002). Figure 2.3 synergizes key concepts and contributions from environmental science and international relations, aiming to improve outcomes for environmental peacebuilding initiatives.

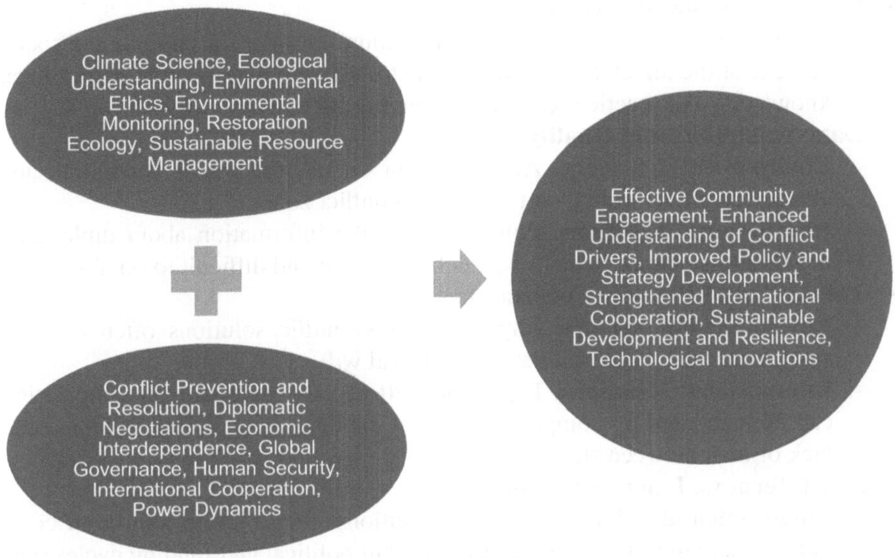

Fig. 2.3 Integrating Disciplinary Insights: Synergizing Environmental Science and International Relations. Note. This figure synergizes key concepts and contributions from environmental science and international relations, aiming to improve outcomes for environmental peacebuilding initiatives

2.7 Challenges of Cross-Disciplinary Approaches to Environmental Peacebuilding

Cross-disciplinary approaches face significant real-world challenges (Matthew et al., 2002). These challenges often arise from the distinct theoretical foundations and disciplinary perspectives inherent to each field. Effectively addressing these issues necessitates targeted capacity building, policy reforms at the disciplinary level, or customized actions tailored to specific situations. There are meaningful opportunities for future research. Among the key challenges are:

- **Complexity of Environmental Issues**
  - Environmental challenges like biodiversity loss, climate change, and pollution are complex and interconnected, making them difficult to address holistically.
- **Cross-Disciplinary Collaboration**
  - Effective communication and collaboration between professionals from environmental science and international relations are crucial but challenging due to cultural resistance and institutional barriers as well as their differing communication platforms and styles, data collection methods, focal points, funding resources, guiding principles, integrative approaches, methodologies, perspectives, and power dynamics.

- **Cultural and Social Factors**
  - Environmental peacebuilding initiatives must consider the cultural and social context of the affected communities, but understanding and integrating local knowledge and practices can be challenging for external actors.
- **Data Availability and Quality**
  - **Environmental Science.** Accurate data on environmental conditions and changes is often limited, particularly in conflict zones.
  - **International Relations.** Political data, like information about diplomatic negotiations or state behaviors, can be sensitive and difficult to obtain.
- **Political and Economic Constraints**
  - **Environmental Science.** Implementing scientific solutions often requires significant financial resources and political will.
  - **International Relations.** Diplomatic efforts and international cooperation can be hampered by competing national interests, geopolitical tensions, and lack of trust between states.
- **Short-Term vs. Long-Term Goals**
  - **Environmental Science.** Scientific solutions often require long-term commitment and monitoring to be effective, but political and funding cycles may prioritize short-term results.
  - **International Relations.** Diplomatic strategies and agreements can be influenced by short-term political agendas, which may not align with the long-term goals of environmental peacebuilding.

2.8 End Notes

Environmental peacebuilding offers a holistic approach to conflict prevention, resolution, and sustainable development by leveraging the power of nature to foster cooperation and resilience. It paves the way for lasting peace and sustainability by addressing the root causes of conflict and meeting communities' needs for a healthy environment and sustainable livelihoods (Conca & Dabelko, 2002; Matthew et al., 2002).

However, cross-disciplinary approaches are essential in environmental peacebuilding, drawing on diverse perspectives and expertise to tackle complex challenges more effectively. This integration fosters innovative, contextually relevant, and sustainable solutions, ensuring initiatives are robust and adaptable to various situations (Bennett & Chapin, 2018; Ostrom, 2009). By collaborating across disciplines, we can create comprehensive strategies that resolve conflicts, promote long-term peace, ensure environmental sustainability, and enhance community well-being.

This précis demonstrated that insights from environmental science and international relations are vital for understanding the complex interplay between environmental factors and conflict dynamics. By integrating these disciplinary perspectives, we can develop effective strategies for managing natural resources, promoting inclusive governance, and enhancing community resilience (Homer-Dixon, 1999; Koubi et al., 2014).

References

Adler, E. (2019). *World ordering: A social theory of cognitive evolution*. Cambridge University Press.

Barnett, J. (2001). *The meaning of environmental security: Ecological politics and policy in the new security era*. Zed Books.

Barnett, J. (2018). *Environmental science and international relations: Comparative approaches and methodologies*. Cambridge University Press.

Begon, M., Townsend, C., & Harper, J. (2006). *Ecology: From individuals to ecosystems* (4th ed.). Blackwell Publishing.

Bennett, E., & Chapin, F. (2018). Interactions between ecology and environmental science. *Environmental Science & Technology, 52*(14), 7632–7641.

Bodansky, D., Brunnée, J., & Hey, E. (Eds.). (2007). *The Oxford handbook of international environmental law*. Oxford University Press.

Callicott, J., & Frodeman, R. (2009). *Encyclopedia of environmental ethics and philosophy* (Vol. 1). Macmillan Reference USA.

Conca, K., & Dabelko, G. (2002). *Environmental peacemaking*. Johns Hopkins University Press.

Conca, K., & Wallace, J. (2009). Environment and peacebuilding in war-torn societies: Lessons from the UN Environment Programme's experience with post-conflict assessment. *Global Governance, 15*(4), 485–504.

Cunningham, W., & Cunningham, M. (2022). *Environmental science: A global concern* (15th ed.). McGraw-Hill Education.

Dresse, A., Fischhendler, I., Nielsen, J., & Zikos, D. (2019). Environmental peacebuilding: Towards a theoretical framework. *Cooperation and Conflict, 54*(1), 99–119.

Dresse, C., Ide, T., & Ide, R. (2018). Environmental peacebuilding: A review of the empirical evidence and future research directions. *Journal of Peace Research, 55*(4), 491–503.

Dunne, T., Kurki, M., & Smith, S. (Eds.). (2021). *International relations theories: Discipline and diversity* (5th ed.). Oxford University Press.

Goldstein, J., & Pevehouse, J. (2017). *International relations* (11th ed.). Pearson.

Heffington, E., Krampe, C., & Scheffran, J. (2019a). Environmental peacebuilding: A critical review of the literature. *Journal of Peace Research, 56*(5), 679–695.

Heffington, S., Oler, A., & Tretler, D. (Eds.). (2019b). *A national security strategy primer*. National Defense University Press.

Homer-Dixon, T. (1999). *Environment, scarcity, and violence*. Princeton University Press.

Ide, T. (2021). Environmental peacebuilding: Moving beyond resolving violence-ridden conflicts to sustaining peace. *World Development, 139*, 105348.

Ikenberry, G. (2018). *A world safe for democracy: Liberal internationalism and the crises of global order*. Yale University Press.

Koubi, V., Spilker, G., Böhmelt, T., & Bernauer, T. (2014). Do natural resources matter for interstate and intrastate armed conflict? *Journal of Peace Research, 51*(2), 227–243.

Krampe, C., & Swain, A. (2024). Environmental peacebuilding and conflict prevention: New insights and approaches. *International Environmental Agreements, 24*(1), 123–145.

Krampe, F., O'Driscoll, D., Johnson, M., Simangan, D., Hegazi, F., & de Coning, C. (2024). Climate change and peacebuilding: Sub-themes of an emerging research agenda. *International Affairs, 100*(3), 1111–1130.

Kydd, A. (2015). *International relations theory*. Cambridge University Press.

Matthew, R., Brown, O., & Jensen, D. (2009). *From conflict to peacebuilding: The role of natural resources and the environment*. United Nations Environment Programme.

Matthew, R., Halle, M., & Switzer, J. (2002). *Conserving the peace: Resources, livelihoods, and security*. International Institute for Sustainable Development.

Meadows, D. (2008). *Thinking in systems: A primer*. Chelsea Green Publishing.

Mearsheimer, J. (2018). *The great delusion: Liberal dreams and international realities*. Yale University Press.

Millennium Ecosystem Assessment. (2005). *Ecosystems and human well-being: Synthesis*. Island Press.

Mingst, K., & Arreguín-Toft, I. (2016). *Essentials of international relations* (7th ed.). W. W. Norton.

National Aeronautics and Space Administration. (n.d.). *Climate change: How do we know?* NASA. https://climate.nasa.gov/evidence/

Odum, E. (1971). *Fundamentals of ecology* (3rd ed.). W.B. Saunders Company.

Ostrom, E. (2009). A general framework for analyzing sustainability of social-ecological systems. *Science, 325*(5939), 419–422.

Peace Direct & Alliance for Peacebuilding. (2020). *Local peacebuilding: What works and why*. Peace Direct.

Reus-Smit, C., & Snidal, D. (Eds.). (2008). *The Oxford handbook of international relations*. Oxford University Press.

Roskin, M., & Cord, R. (2019). *Political science: An introduction* (14th ed.). Pearson.

Rupert, M., & Smith, H. (2016). *Historical materialism and globalization*. Routledge.

Scoones, I. (1998). *Sustainable rural livelihoods: A framework for analysis*. IDS Working Paper 72.

Serrat, O. (2017). The sustainable livelihoods approach. In *Knowledge solutions: Tools, methods, and approaches to drive organizational performance* (pp. 21–26). Springer.

Serrat, O. (2025a). Environmental peacebuilding: Nurturing hope with analytical persuasion. In *Anthropogenic solutions for climate change: Achieving environmental peace* (pp. 165–170). Springer.

Serrat, O. (2025b). Safeguarding Lake Chad. Nurturing hope with analytical persuasion. In: *Anthropogenic solutions for climate change: Achieving environmental peace* (pp. 243–257).

Serrat, O. (2025c). Safeguarding the mother of waters: Environmental peacebuilding in the Mekong River Basin. In: *Anthropogenic solutions for climate change: Achieving environmental peace* (pp. 229–239).

Slingerland, E. (2008). *What science offers the humanities: Integrating body and culture*. Cambridge University Press.

Sommer, U., & Fassbender, F. (2024). Environmental peacebuilding: Moving beyond resolving violence-ridden conflicts to sustaining peace. *World Development, 139*, 105348.

True, J. (2018). *The political economy of violence against women*. Oxford University Press.

United Nations Environment Programme. (2023). *Environmental cooperation and peacebuilding: A guide for practitioners*. United Nations Environment Programme.

United Nations. (n.d.). *Sustainable development goals: What are the sustainable development goals?* https://sdgs.un.org/goals

Withgott, J., & Laposata, M. (2012). *Essential environment: The science behind the stories* (4th ed.). Pearson.

Chapter 3
Bridging the Gap Synergizing Anthropology, Economics, History, and Sociology for Environmental Peacebuilding

Abstract Environmental peacebuilding integrates the management of natural resources and environmental challenges not only for conflict mitigation but also as a proactive strategy to prevent disputes before they arise. This précis examines how anthropology, economics, history, and sociology contribute to environmental peacebuilding, demonstrating how insights from these disciplines can inform effective policies and strategies. By outlining the background, research problem, objectives, sub-questions, and literature review for follow-on research, the précis explores how these fields might synergize to foster sustainability, cooperation, and lasting peace.

3.1 Introduction

Imagine a world where nature's elements—air, minerals, soil, water, and other resources—serve as foundational pillars for peace (Conca & Dabelko, 2002). This transformative approach, known as environmental peacebuilding, confronts shared environmental challenges directly (United Nations Environment Programme, 2023). In essence, it leverages cooperation over natural resources to promote and sustain peace. To Dresse et al. (2019), environmental peacebuilding is "the process through which environmental challenges shared by the (former) parties to a violent conflict are turned into opportunities to build lasting cooperation and peace" (p. 104). Definitions advanced by Brown and Nicolucci-Altman (2022) and the United Nations Environment Programme (2024) emphasize the dual potential of environmental peacebuilding—not only facilitating post-conflict recovery but also preempting future conflict, thereby broadening the scope of possible initiatives.

By fostering cooperation, trust, and resilience, environmental peacebuilding tackles root causes of conflict, such as competition over natural resources, environmental degradation, and resource scarcity (Homer-Dixon, 1999).[1] This holistic

[1] Based on geographic location, societal priorities, and ecological challenges, examples of environmental peacebuilding initiatives include climate change adaptation projects, collaborative water management, community-based natural resource management, cross-border wildlife conservation,

O. Serrat, *Advancing Environmental Peacebuilding*, SpringerBriefs in Political Science, https://doi.org/10.1007/978-981-92-4028-9_3

approach lays the foundation for enduring peace and long-term stability, ensuring that communities can thrive with sustainable livelihoods and a healthy environment (Dresse et al., 2018; Heffington et al., 2019a; Krampe et al., 2024; Scoones, 1998; Serrat, 2017; United Nations Environment Programme, 2023). Case Study 1 and Case Study 2 illustrate successful environmental peacebuilding in action—each offering a concrete example that helps illuminate the concept both on the ground and in diplomacy.

Case Study 1 The Good Water Neighbors ProjectOverview

- The Good Water Neighbors Project initiative led by EcoPeace/Friends of the Earth Middle East was launched in 2001, working until 2005 with 11 Israeli, Palestinian, and Jordanian communities and expanding in Phase II, 2005–2008, to 17 communities. The project brought together communities to collaboratively manage scarce water resources in Wadi Fukin and Tzur Hadassah. By establishing joint water management strategies and local monitoring systems, the project not only improved water infrastructure but also fostered trust, dialogue, and reconciliation among historically divided groups. The project demonstrated how communities with a long history of conflict can transform a shared environmental challenge into an opportunity for peace.

Key Insights

- **Social Cohesion:** Through regular inter-community meetings and collaborative problem-solving, participants built personal relationships that extended beyond water management.
- **Concrete Impact:** The project provided tangible benefits by stabilizing water supply and reducing tensions over a limited resource.
- **Model for Replication:** This case study neatly encapsulates how environmental challenges can be addressed with collaborative local action that contributes to broader peacebuilding efforts.

Source: Adapted from EcoPeace/Friends of the Earth Middle East (2008). *Environmental peacebuilding theory and practice: A case study of the Good Water Neighbors Project and in-depth analysis of the Wadi Fukin/Tzur Hadassah communities*. https://ecopeaceme.org/wp-content/uploads/2022/03/environmental-peacebuilding-theory-and-practice-a-case-study-of-the-good--water-neighbours-projectand-in-depth-analysis-of-the-wadi-fukin-tzur-hadassah-communities.pdf

disaster risk reduction initiatives, forest conservation and management, integrated coastal management programs, joint environmental monitoring, marine conservation agreements, partnerships for biodiversity conservation, reforestation and land restoration projects, renewable energy projects, sustainable agriculture programs, transboundary air pollution agreements, and transboundary protected areas.

Case Study 2 Diplomatic Mediation in Environmental Cooperation

Overview

- At the international level, environmental peacebuilding often takes the form of diplomatic mediation where shared natural resources act as a bridge between conflicting parties. An illustrative example is the involvement of the United Nations Environment Programme (UNEP) in transboundary conflicts. UNEP mediates disputes by providing technical expertise and acting as a neutral party in negotiations over shared resources—such as water bodies, forests, or mineral deposits. These efforts help transform potential flashpoints into platforms for cooperative management and sustainable policy development.

Key Insights

- **Technical Mediation:** UNEP's role as a neutral expert ensures that technical know-how on natural resources informs diplomatic negotiations.
- **Confidence Building:** By grounding discussions in environmental science and sustainable management practices, diplomatic mediators build a foundation of trust and shared purpose among the involved nations.
- **Policy Integration:** This approach not only addresses immediate resource conflicts but also supports longer-term goals like sustainable development and conflict prevention.

Source: Adapted from United Nations Environment Programme (2025). *Environmental cooperation and peacebuilding.* https://www.unep.org/topics/disasters-and-conflicts/environment-security/environmental-cooperation-and-peacebuilding

Successful environmental peacebuilding efforts inevitably depend on inclusive stakeholder engagement, strong collaboration, innovative strategies, ongoing adaptability, and comprehensive evaluation of lessons learned (Scoones, 1998; Serrat, 2025a, 2025b). Each initiative must carefully analyze the strategic landscape, establish clear objectives, identify the necessary resources to achieve them, devise effective ways to utilize these resources, and assess potential costs and risks (Heffington et al., 2019b).

However, despite their promise, many environmental peacebuilding efforts are hindered by fragmentation, stemming from the absence of an integrated framework that unites diverse disciplinary insights (Serrat, 2025c). To address this, success in environmental peacebuilding requires a holistic and interdisciplinary strategy, leveraging cultural, economic, historical, and sociological perspectives. Such an approach avoids the pitfalls of reductionism, which often oversimplifies complex problems when analyzed in isolation (Sommer & Fassbender, 2024). While specialized knowledge is valuable, it must complement rather than dominate other perspectives. Rooted in pragmatism, the most profound understanding arises at the intersections of various disciplines, where methods like abduction, deduction, and induction offer deeper insights (Slingerland, 2008). Philosophical principles like positivism, pragmatism, and realism play a crucial role in guiding this integrative process, with pragmatism often proving particularly effective. Environmental peacebuilding, like many other fields, thrives when disciplines collaborate rather than compete. By fostering such synergy, it becomes possible to develop solutions that are not only comprehensive but also sustainable and transformative, paving the way for lasting peace and stability.

At least in theory, environmental peacebuilding is inherently integrative, drawing insights from both the natural and social sciences to understand and manage the intersections of environmental challenges and conflict dynamics. For instance, enhanced academic rigor suggests the combined efforts of anthropology, economics, history, and sociology, as examples discussed in this précis, can significantly upgrade the effectiveness of environmental peacebuilding.[2] While Case Studies 1 and 2 exemplify effective environmental peacebuilding, contrasting outcomes in Case Studies 3 and 4 highlight the pitfalls of inadequately integrated approaches.

Case Study 3 The Great Limpopo Transfrontier Park

Originally conceived as a collaborative effort to promote cross-border environmental conservation between Mozambique, South Africa, and Zimbabwe, the Great Limpopo Transfrontier Park initiative was intended to foster peace through shared natural resource management. However, the creation of the park led to the displacement of local communities who had long depended on the land for their livelihoods. The failure to adequately address the rights and needs of these communities not only undermined the social fabric of the affected areas but also highlighted the risk that projects aimed at mitigating conflict can inadvertently contribute to new forms of injustice and instability.

Note. Integrating anthropological insights on cultural ties, economic evaluations for equitable livelihoods, historical analyses to acknowledge colonial legacies, and sociological studies of community networks could have guided more sensitive and inclusive planning to prevent harmful displacement, while further contributions from other disciplines could have enhanced the initiative even more.

Source. Adapted from Lunstrum (2010). Reconstructing history, grounding claims to space: History, memory, and displacement in the Great Limpopo Transfrontier Park. *South African Geographical Journal*, *92*(2), 129–143.

[2] In addition to anthropology, economics, history, and sociology, the core disciplines shaping environmental peacebuilding include ecology; environmental science; human geography; international relations; law; peace and conflict studies; populism studies; and sustainable development. Depending on the specific context, other relevant fields may encompass environmental education, environmental engineering, environmental ethics, environmental psychology, geology, hydrology, public health, sociology, and urban planning.

Case Study 4 The Israeli-Jordanian-Palestinian Water Channel Project

The Israeli-Jordanian-Palestinian Water Channel Project, a multi-party initiative, was designed to establish a channel from the Red Sea to the Dead Sea, with the goal of enhancing water and energy security for all involved. Despite its promising objectives, the project ultimately faltered because it could not overcome the deep-seated inequities in water access and decision-making between the parties, particularly the persistent imbalance that disadvantaged Palestinian communities. Faced with these unresolved political and socio-economic challenges, the project was cancelled in 2021.

Note. Combining anthropological understandings of water's cultural significance, economic assessments of fair resource allocation, historical context of longstanding water disputes, and sociological evaluations of power dynamics might have fostered a more equitable and sustainable collaborative framework, while further contributions from other disciplines could have enhanced the initiative even more.

Source. Adapted from Ide (2020, January 9). *Beware the dark side of environmental peacebuilding*. Climate Diplomacy. https://climate-diplomacy.org/magazine/conflict/beware-dark-side-environmental-peacebuilding

3.2 Background

3.2.1 Emergence and Evolution of Environmental Peacebuilding

Environmental peacebuilding emerged in the 1980s as practitioners and scholars recognized that competition over natural resources, environmental degradation, and resource scarcity can intensify social tensions and spark conflict (Homer-Dixon, 1999). Environmental peacebuilding holds particular importance in our increasingly interconnected world, as environmental challenges (e.g., climate change, deforestation, water scarcity) pose significant threats to both local and global stability (Heffington et al., 2019b). For example, overfishing in coastal areas threatens livelihoods, deforestation in the Amazon disrupts biodiversity and local communities while exacerbating global climate change, and the worldwide impacts of climate change lead to displacement, food insecurity, and increased tensions. These examples underscore the urgent need for an integrated approach to environmental peacebuilding (Dresse et al., 2018).

3.2.2 Environmental Influences on Conflict: A Complex Interplay

At first, environmental peacebuilding was concerned only with the role of environmental factors in shaping conflict dynamics. However, the field has since evolved to explore their complex interplay, integrating strategies that leverage environmental management to foster peace (Krampe et al., 2024; Krampe & Swain, 2024). Tackling the complex challenges of environmental conflicts demands perspectives from cultural, ecological, economic, political, and social spheres.[3] Cross-disciplinarity enables a thorough understanding of environmental influences and conflict dynamics. Drawing on the strengths of diverse fields, environmental peacebuilding can construct a more comprehensive framework to address environmental drivers of conflict and support long-term peace and stability.

3.2.3 Core Goals of Environmental Peacebuilding

Effective environmental peacebuilding initiatives must be tailored to the unique environmental, political, and social contexts of each locale. Accordingly, these initiatives pursue diverse objectives such as fostering trust and cooperation through collaborative environmental projects, managing shared natural resources to reduce tensions and prevent conflicts, ensuring fair and equitable treatment through environmental justice, and engaging diverse stakeholders in inclusive governance (Krampe & Swain, 2024). Additional objectives include strengthening community resilience and promoting sustainable development for both environmental and local well-being (Matthew et al., 2009; Conca & Dabelko, 2002; Conca & Wallace, 2009). These objectives can overlap, mutually reinforcing one another to create a more resilient framework.

[3] For example, initiatives to curb deforestation in the Amazon often encounter resistance from competing conservation and economic interests (e.g., logging) that bypass legislation. Collaborative efforts involving indigenous leaders, ecologists, economists, and sociologists—coupled with real-time satellite monitoring and stricter penalties for violations—could foster sustainable strategies to protect ecosystems. In Colombia, post-conflict initiatives to restore degraded lands and integrate former combatants into conservation work face logistical and social hurdles. The involvement of environmental scientists, social workers, and conflict resolution specialists could invigorate these efforts and enhance their long-term success. In the Lake Chad region, environmental degradation and water scarcity have intensified conflicts among communities reliant on the lake, spanning four countries (Serrat, 2025d). A multidisciplinary approach, integrating diplomats, climate scientists, agricultural experts, and conflict resolution specialists, could address environmental and social dimensions, paving the way for resilience. Similarly, in the Mekong River basin, dam construction has exacerbated international disputes over water rights, ecological harm, and impacts on fisheries and agriculture (Serrat, 2025e). Cross-disciplinary teams of hydrologists, environmental scientists, diplomats, and local representatives could bolster transboundary water governance. In all these cases, stronger political will remains key to transforming challenges into opportunities.

3.3 Research Question

According to the Global Atlas of Environmental Justice (https://ejatlas.org/), more than 4200 environmental conflicts were documented worldwide as of early 2024. Although academic interest in environmental peacebuilding is growing, Google Scholar only lists approximately 2400 articles to date, with over 90% published after 2010 and 73% appearing in the last five years. Moreover, this surge in scholarly output has not translated into a unified body of knowledge. Instead, the field remains fragmented across disciplinary silos, leading to a disjointed understanding of complex environmental challenges. The pressing need for integration is underscored by the complementary insights offered by different academic perspectives. Anthropology contributes rich cultural narratives and indigenous wisdom; economics dissects resource allocation and market dynamics; history provides a long-term context to current conflicts by unveiling legacies and patterns; and sociology sheds light on the social structures and the forces that shape community interactions. Despite these significant contributions, the absence of an integrated framework means that current efforts in environmental peacebuilding are limited by academic fragmentation rather than by inadequate financial resources. Against this backdrop, the research question guiding this précis is: How can anthropology, economics, history, and sociology contribute to environmental peacebuilding initiatives?

This précis explores how these disciplines address issues from different perspectives, uniquely illuminate the research question, provide answers, and diverge in their investigative approaches. Finally, based on the foregoing analyses, this précis suggests a plan for follow-on research to integrate the perspectives of anthropology, economics, history, and sociology and make further progress on the research question.

3.4 Theoretical Foundations of Anthropology, Economics, History, and Sociology

To appreciate how anthropology, economics, history, and sociology contribute to environmental peacebuilding, one must examine their theoretical foundations (Bennett & Chapin, 2018; Sommer & Fassbender, 2024). Understanding the key concepts, principles, and perspectives of distinct disciplines enables us to identify effective strategies for fostering environmental peace and stability. Why? Because by delving into their theoretical underpinnings, we can uncover synergies and complementarities to address complex environmental challenges and—hopefully—promote sustainable, long-lasting peace.

3.4.1 *Key Concepts, Principles, and Perspectives of Anthropology*

Anthropology explores human societies, cultures, and biological origins, offering insights into the diverse and intricate fabric of human experience (Kottak, 2013; Ember & Ember, 2014). It rests on principles from holistic inquiry, cultural relativism, and ethnographic methodology to address complex questions about identity, social organization, and communication. Key concepts include cultural relativism, which advocates understanding practices within their own contexts (Benedict, 1934); human evolution, which examines the biological and genetic roots shaping our species (Darwin, 1859); kinship and social structure, which decipher patterns of familial and communal relationships (Lévi-Strauss, 1963); and linguistic anthropology, which interprets language as a pivotal medium for expressing cultural meanings (Whorf, 1956). Beyond these foundational pillars, contemporary anthropology is expanding its scope to integrate digital ethnography, critical theory, and postcolonial perspectives—thereby addressing the nuances of globalization and modern social transformations.

3.4.2 *Key Concepts, Principles, and Perspectives of Economics*

Economics is rooted in understanding how individuals, firms, and societies allocate scarce resources to satisfy unlimited wants, a challenge that shapes decision-making at every level (Mankiw, 2021; Krugman & Wells, 2018). Drawing on both microeconomic and macroeconomic frameworks, it dissects behaviors—from the everyday choices of consumers and producers to the large-scale dynamics that drive national growth. Key concepts include supply and demand, which articulate how buyers' desires and sellers' offerings converge to set market prices (Smith, 1776; Marshall, 1890), and opportunity cost, which reminds us that every decision involves sacrificing an alternative (Samuelson & Nordhaus, 1992). Beyond these foundational ideas, economics leverages principles of market equilibrium and incentives to explain how markets function and adjust over time. Modern perspectives increasingly incorporate insights from behavioral economics to challenge the classic assumption of complete rationality (Thaler, 2015), while also exploring market failures—such as externalities and information asymmetries—that reveal where and how interventions may boost efficiency. Together, these concepts and perspectives enable a nuanced analysis of economic policies, global trade, income distribution, and the overall health of an economy.

3.4.3 Key Concepts, Principles, and Perspectives of History

History is founded in the study of past events, cultural shifts, and the complex narratives that influence our understanding of the present (Carr, 1961; Tosh, 2015). It rests on principles such as critical inquiry, rigorous source analysis, and interpretative pluralism, enabling historians to unravel the interconnected causes and consequences of human actions. Key concepts include historical significance, which examines why certain events are commemorated or relegated to obscurity; source criticism, which emphasizes the careful evaluation of both primary and secondary evidence to construct trustworthy narratives (Bloch, 1987); causality, which explores the dynamic interplay between events that trigger broader social, economic, or political shifts; and perspective, which underscores the diverse viewpoints that contribute to a more nuanced understanding of the past (Seixas & Morton, 2013). Modern historiography further embraces digital methodologies, interdisciplinary approaches, and ethical considerations, ensuring that historical inquiry remains a dynamic, reflective dialogue between past and present.

3.4.4 Key Concepts, Principles, and Perspectives of Sociology

Sociology investigates the frameworks of human relationships, institutional forces, and cultural practices that mold society (Giddens, 2009; Ritzer, 2011). It rests on principles of empirical inquiry, critical analysis, and social theory to explain how both micro-level interactions and macro-level structures influence behavior Fundamental concepts include social structure, viz., the organized patterns of relationships and institutions that shape society (Durkheim, 1997); socialization, the continuous process through which individuals absorb cultural norms, values, and roles (Parsons, 1951); and symbolic interactionism, a perspective that explores how individuals construct and interpret shared meanings in daily interactions (Blumer, 1969). Additionally, conflict theory provides a critical lens to understand power dynamics and social inequalities, drawing on foundational analyses of class struggle (Marx, 1990). Modern sociology also embraces perspectives from digital sociology, cultural studies, and postmodern thought to further unpack the complexities of contemporary social life.

3.5 Environmental Peacebuilding: Disciplinary Insights from Anthropology, Economics, History, and Sociology

3.5.1 Anthropological Perspectives on Environmental Peacebuilding

Anthropological inquiry has long explored how cultural narratives, indigenous knowledge, and historical memory shape human relationships with the natural world. By delving into the lived experiences of communities and their ritual, symbolic, and ethical dimensions toward nature, anthropology reveals that environmental conflict is as much about contested cultural meanings as it is about resource scarcity (Geertz, 1973; Escobar, 1999). This perspective illuminates the ways diverse societies interpret and respond to environmental pressures, highlighting pathways for reconciling differences through culturally sensitive dialogue and collective stewardship. Anthropology plays a vital role in environmental peacebuilding by emphasizing the socio-cultural dimensions that are central to resource management and conflict resolution. Key contributions include ethnographic research, which uncovers how local practices and oral traditions inform sustainable resource usage; cultural analysis, which deciphers the symbolic value attached to landscapes and environmental rituals; historical contextualization, which traces the origins of environmental grievances; and participatory approaches that empower communities to shape peacebuilding initiatives. Together, these insights foster strategies that are not only ecologically sound but also imbued with the cultural wisdom and resilience necessary for lasting peace (Ingold, 2000).

3.5.2 Economic Perspectives on Environmental Peacebuilding

Economics studies resource distribution, market interactions, and the role of policy in guiding economic behavior (Krugman & Wells, 2018; Mankiw, 2021). In environmental peacebuilding, economic analysis reveals how scarcity, opportunity costs, and externalities drive conflicts over natural resources. By quantifying the trade-offs inherent in decisions regarding resource use and conservation, economic insights help frame policies that not only promote sustainable development but also mitigate the risks of unrest and instability (Samuelson & Nordhaus, 1992).

Beyond traditional market analysis, economics enriches environmental peacebuilding by employing tools such as cost-benefit analysis, game theory, and behavioral economics. These approaches reveal how strategic interventions—ranging from tax incentives and sustainable subsidies to private-public partnerships—can align economic interests with environmental stewardship. By transforming competitive pressures into collaborative ventures, economic strategies enable communities and decision-makers to foster equitable resource governance and create conditions that support long-term peace and resilience (Thaler, 2015).

3.5.3 *Historical Perspectives on Environmental Peacebuilding*

History examines past events, narratives, and collective memory to shed light on the roots of environmental conflicts and the evolution of resource use. By analyzing archival records, oral histories, and material culture, historical inquiry reveals long-term patterns—such as the impact of colonial exploitation, industrialization, and shifting political boundaries—that have shaped the dynamics between societies and their environments (Carr, 1961; Tosh, 2015).

Historical insights uniquely contribute to environmental peacebuilding by contextualizing current disputes within enduring legacies of injustice and cultural memory. Key contributions include revealing the origins of resource-based conflicts, tracing the transformation of environmental policies across eras, and identifying recurring themes in human–environment interactions. This temporal perspective highlights lessons from past mismanagement and exploitation and guides strategies for reconciliation (Bloch, 1987; Seixas & Morton, 2013).

3.5.4 *Sociological Perspectives on Environmental Peacebuilding*

Sociology illuminates the intricate interplay between social structures, collective identities, and environmental dynamics. By examining how power relations, class disparities, and community networks shape both resource distribution and conflict dynamics, sociology provides vital insights into the social roots of environmental unrest. Techniques such as social network analysis and qualitative research reveal how systemic inequalities can intensify environmental degradation and spur contested claims over natural resources (Bullard, 2000; Pellow, 2000). Moreover, sociological inquiry emphasizes the role of social movements, cultural norms, and participatory processes in mobilizing communities toward sustainable management and peaceful conflict resolution. By elucidating the social foundations of environmental justice and resilience, this perspective informs policies that foster inclusive decision-making, social cohesion, and grassroots diplomacy—key elements for building lasting peace in environmentally stressed regions (Castells, 2012).

3.5.5 *Methodological Insights from Anthropology, Economics, History, and Sociology*

Given their distinct disciplinary origins, anthropology, economics, history, and sociology adopt diverse investigative methods. These approaches shape how each field explores and interprets the world, as illustrated below.

3.5.5.1 The Practice of Anthropology

Anthropology employs immersive fieldwork, ethnographic studies, and interpretative methods to delve into cultural practices. It emphasizes holistic understandings of traditions, rituals, and meaning-making processes across diverse communities, revealing how shared symbols and narratives shape identities and social relationships (Geertz, 1973; Escobar, 1999).

3.5.5.2 The Practice of Economics

Economics analyzes the allocation of scarce resources through mathematical modeling, quantitative analysis, and empirical methods, unveiling the complexities of market behavior and incentive structures. It examines both micro-level decisions and macroeconomic policies to understand trade-offs and forecast outcomes, thereby providing frameworks for efficient resource management and policy formulation (Mankiw, 2021; Samuelson & Nordhaus, 1992).

3.5.5.3 The Practice of History

History investigates past events through archival research, source analysis, and narrative reconstruction. Its practice entails analyzing long-term patterns of social, political, and cultural change to contextualize contemporary phenomena. By tracing legacies of conflict, injustice, and transformation, historical inquiry unearths the roots of modern challenges and informs strategies geared toward reconciliation and progress (Carr, 1961; Tosh, 2015).

3.5.5.4 The Practice of Sociology

Sociology leverages qualitative and quantitative methods to explore social structures, inequality, and community dynamics. Through surveys, case studies, and in-depth interviews, sociologists explore how power relations and cultural norms influence collective behavior and institutional development, providing critical insights into strategies for promoting social cohesion and inclusive governance (Durkheim, 1997; Giddens, 2009).

3.5.5.5 Contrasting Investigative Methods: Anthropology, Economics, History, and Sociology

The key differences among anthropology, economics, history, and sociology are evident in three central aspects: their research methods, the scope of analysis, and the way they establish context.

In terms of research methods, anthropology and history favor rich qualitative approaches. They use ethnographic studies and archival research to capture the subtleties of cultural practices and historical shifts. In contrast, economics leans on a quantitative framework—relying on mathematical models and statistical analysis to translate individual behaviors into broad market trends. Sociology strikes a balance by combining interviews and case studies with surveys, offering a well-rounded view of social structures and power dynamics.

When it comes to the depth versus breadth of analysis, anthropology and history prioritize deep, detailed investigations into human experience and cultural evolution. Economics, however, streamlines individual complexities into broader models that reveal overall market trends and policy impacts. Sociology merges these approaches by blending detailed case studies with wider surveys, thereby capturing insights at both the micro and macro levels.

Differences are also evident in how each discipline presents findings. Anthropology and history use narrative techniques to depict cultural, historical, and social settings, revealing the meanings behind human behavior. In contrast, economics tends to set aside detailed context in favor of building general models—an approach that supports clear predictions but can overlook situational detail. Meanwhile, sociology blends these strategies by considering both contextual influences and the broader patterns that emerge across various social settings (Table 3.1).

By contrasting their respective investigative methods and areas of emphasis, we not only appreciate each discipline's unique perspective but can also uncover opportunities for interdisciplinary collaboration. A comparative approach underscores that while each discipline operates with its own distinct toolkit, their methodologies can be complementary. Embracing these differences enriches our overall understanding of human behavior, societal dynamics, and the challenges we face today.

Table 3.1 A Comparative Overview of Anthropology, Economics, History, and Sociology

Discipline	Primary Methodologies	Emphasis	Core Contribution
Anthropology	Ethnography, immersive fieldwork, participant observation	Rich contextual and qualitative insight	Anthropology unveils lived experiences and cultural meanings
Economics	Mathematical modeling, quantitative analysis, empirical experiments	Abstraction, generalizability, prediction	Economics predicts trends and quantifies trade-offs
History	Archival research, source criticism, narrative reconstruction	Temporal context and qualitative depth	History reveals long-term patterns and transformative eras
Sociology	Surveys, case studies, mixed qualitative and quantitative methods	Integration of both qualitative and quantitative approaches	Sociology balances case insight with broad social trends

3.6 Practical Implications of Integrating Disciplinary Insights

Integrating insights from anthropology, economics, history, and sociology into environmental peacebuilding initiatives offers a multifaceted approach to resolving conflicts rooted in environmental change (Conca & Dabelko, 2002). Each discipline contributes a distinct analytical lens: anthropology reveals rich cultural narratives, indigenous knowledges, and local values that shape human–nature interactions; economics provides models and quantitative analyses to understand resource scarcity, opportunity costs, and market incentives; history traces the legacies of past conflicts and institutional trajectories that continue to influence present-day disputes; and sociology reveals the social structures, power dynamics, and community networks that shape collective action and foster resilience.

These combined perspectives have several practical implications. First, tailored community engagement becomes possible when culturally informed approaches are integrated with economic assessments and historical context, fostering participatory decision-making and equitable resource management. Second, policy and strategy development benefit from a holistic analysis that blends statistical rigor with qualitative insights, helping to craft interventions that are both sustainable and sensitive to longstanding grievances. Third, the interdisciplinary approach enhances coordination among local, national, and international stakeholders by bridging technical, historical, and sociocultural dimensions, which is critical for designing adaptive and resilient peacebuilding initiatives. Different disciplines working together help environmental peacebuilding address the root causes of conflict, leading to more stable, inclusive, and longer-lasting solutions (Conca & Dabelko, 2002; Millennium Ecosystem Assessment, 2005).

3.7 Limitations

This précis advocated for the integration of disciplinary insights into environmental peacebuilding initiatives. Nevertheless, there are considerable real-world challenges inherent in cross-disciplinary approaches, many of which stem from the theoretical foundation and disciplinary perspective of each field. Addressing these challenges requires targeted capacity building, policy improvements at the disciplinary level, or tailored actions on a case-by-case basis. Among the most notable challenges are cultural resistance and institutional barriers. Furthermore, it is essential to reconcile differences in communication platforms and styles, data collection methods, focal areas, funding sources, guiding principles, integrative strategies, methodologies, perspectives, and power dynamics. While these obstacles pose significant difficulties, they also present valuable opportunities for transformative research—bridging epistemological divides, refining interdisciplinary methodologies, and driving more effective and integrative environmental peacebuilding initiatives (Pellini & Serrat, 2017).

3.8 Discussion

Environmental peacebuilding takes a comprehensive approach to preventing and resolving conflicts while promoting sustainable development, harnessing the power of nature to strengthen cooperation and resilience. It paves the way for lasting peace and sustainability by addressing the root causes of conflict and meeting communities' needs for a healthy environment and sustainable livelihoods (Conca & Dabelko, 2002; Matthew et al., 2002).

Cross-disciplinary approaches are crucial in environmental peacebuilding, as they integrate diverse perspectives and expertise to address complex challenges with greater efficiency. This integration promotes the development of innovative, context-aware, and sustainable solutions, ensuring that initiatives remain resilient and adaptable across diverse contexts (Bennett & Chapin, 2018; Ostrom, 2009). This précis demonstrated that insights from anthropology, economics, history, and sociology are vital for understanding the complex interplay between environmental factors and conflict dynamics. By integrating these disciplinary perspectives, and addressing the challenges of cross-disciplinary approaches, we can develop effective strategies for managing natural resources, promoting inclusive governance, and enhancing community resilience (Homer-Dixon, 1999; Koubi et al., 2014).

3.9 The Proposed Follow-On Research

By exclusively drawing on secondary sources and published literature, the proposed follow-on research would highlight how the unique vocabulary and theoretical constructs of anthropology, economics, history, and sociology can be aligned into a common framework. The proposed research would demonstrate that, despite interdisciplinary divergences and disputes, a coherent language can emerge to address environmental conflicts. The proposed research would systematically review theoretical contributions, identify common themes, and propose an integrative model that facilitates cross-disciplinary dialogue and informs policy recommendations—all through a rigorous desk-based approach.

3.9.1 Background and Motivation

Rising environmental challenges—such as resource depletion, climate change, and ecosystem degradation—often catalyze conflict. However, each discipline interprets these challenges differently: anthropology emphasizes cultural narratives and indigenous knowledge; economics focuses on resource allocation and market dynamics; history examines the legacies and lessons of past environmental conflicts; sociology investigates the social structures and networks that influence community resilience. The proposed research would contend that these distinct intellectual traditions can be reconciled to offer a more holistic approach to environmental peacebuilding.

3.9.2 Research Problem

Disciplinary silos often lead to fragmented understanding and policy prescriptions. To improve environmental peacebuilding, the challenge is to establish a coherent framework where the conceptual languages of anthropology, economics, history, and sociology converge. Building on the research question for this précis, the central inquiry would be: How can the conceptual and methodological vocabularies of anthropology, economics, history, and sociology be reconciled to establish a common framework for environmental peacebuilding?

3.9.3 Research Objectives

Building on this précis. The proposed research would be guided by three interrelated objectives. The first objective would be to identify and document the key conceptual frameworks and terminologies used by anthropology, economics, history, and sociology in addressing environmental conflict and peace. The second objective would be to analyze the divergences and intersecting themes between these frameworks. The third objective would be to synthesize an interdisciplinary model that aligns these distinct perspectives.

3.9.4 Research Sub-Questions

Focusing on the central inquiry specified in the statement of the research problem, the proposed research would aim to answer three sub-questions: (1) What are the central constructs of environmental peacebuilding from the perspectives of anthropology, economics, history, and sociology? (2) How do these constructs converge or diverge, and what common ground exists? (3) What integrative framework can be formulated to bridge these disciplinary discussions?

3.9.5 Literature Review

3.9.5.1 Scope and Approach

Toward the central inquiry, research objectives, and research sub-questions, a systematic review of academic literature would be conducted to map out how each discipline approaches environmental peacebuilding:

- **Anthropology:** The proposed research would focus on ethnographic studies and cultural analyses that discuss indigenous resource management, local conflict narratives, and traditional methods of conflict resolution.
- **Economics:** The proposed research would examine models of resource allocation, market failures, public goods theory, and incentive-based approaches that address environmental sustainability.
- **History:** The proposed research would analyze historical case studies documenting environmental conflicts and subsequent cooperative transformations, highlighting continuity and change over time.
- **Sociology:** The proposed research would investigate research on social networks, collective identity, and power dynamics that elucidate the role of social structures in mediating conflict and fostering peace.

The proposed research would yield a comparative matrix of key terms and concepts in anthropology, economics, history, and sociology, illuminating both the disciplinary divergences and the overlapping themes crucial for building a shared language.

3.9.5.2 Methodology

Without incorporating fieldwork, the proposed research would rely on desk-based methods and secondary data sources. It would undertake a systematic literature review, perform a meta-synthesis and comparative analysis, and develop an integrative conceptual framework to unify concepts and insights across disciplines. The systematic literature review would involve collecting and analyzing peer-reviewed journal articles, scholarly books, and reputable reports from each relevant field, utilizing academic databases to ensure comprehensive coverage of pertinent literature. The meta-synthesis and comparative analysis would code the literature to extract central themes, terminologies, and theoretical constructs, culminating in a conceptual matrix that cross-references the contributions and lexicons of each discipline. Finally, the integrative framework would identify "bridge concepts" serving as translation mechanisms across diverse disciplinary lexicons, and propose a unified framework for environmental peacebuilding. This framework would harmonize insights drawn from anthropology, economics, history, and sociology, fostering interdisciplinary collaboration and understanding.

3.9.6 Preliminary Conclusions and Future Research

The proposed research would highlight how it bridges the disparate terminologies and concepts of anthropology, economics, history, and sociology to promote an integrated approach to environmental peacebuilding. It would stress that aligning these disciplinary perspectives could lead to more informed and effective policymaking. While acknowledging that reliance on secondary literature may constrain

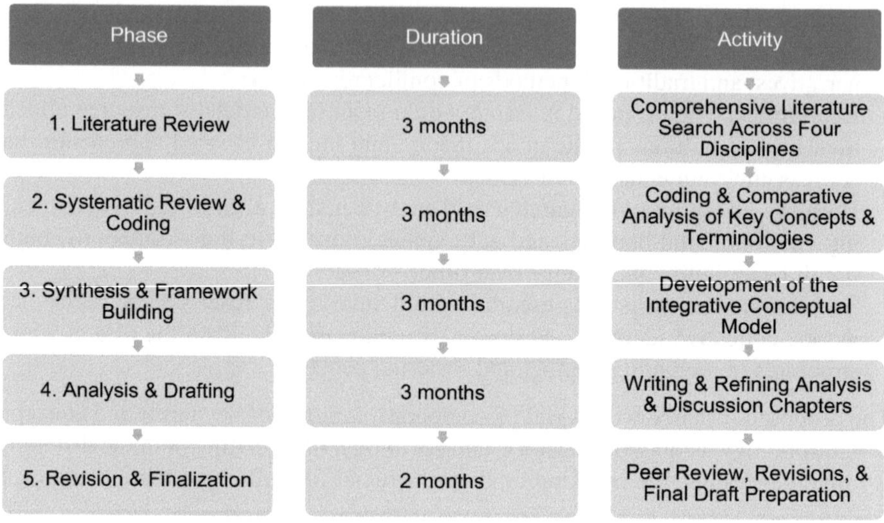

Fig. 3.1 Timeline and Activities for the Proposed Study

empirical insights, the proposed research would argue that its meta-analytical approach provides a strong theoretical foundation for future empirical studies. Furthermore, it would propose that subsequent research could test the integrative framework through targeted case studies or mixed-methods research when circumstances allow for external validation. Future studies could also examine additional intersections, such as digital data analysis or political science, to further enhance the framework's applicability.

3.9.7 Timeline and Activities for the Proposed Research

Figure 3.1 presents a structured timeline outlining the sequence of activities planned for the proposed research. It visually maps each phase—from preliminary preparation through data collection, analysis, and final reporting—highlighting how tasks would progress across the study period.

3.9.8 Resources Needed

The proposed research would require access to databases of academic journals. It would leverage tools for managing and coding qualitative data (e.g., ATLAS.ti, NVivo). Software for creating conceptual models and visual matrices may also be needed.

3.9.9 Broader Impact and Future Directions

The proposed research would underscore the potential for interdisciplinary dialogue to improve our understanding of environmental peacebuilding. By harmonizing the languages of anthropology, economics, history, and sociology, the proposed research would create a pathway for more coherent theoretical frameworks and policy strategies. The integrative framework would not only enrich academic discourse but also promote actionable insights that could inform peacebuilding initiatives. Future directions could involve leveraging the methodologies of digital humanities to enhance interdisciplinary dialogue, paving the way for a range of practical applications.

References

Benedict, R. (1934). *Patterns of culture*. Houghton Mifflin Harcourt.

Bennett, E., & Chapin, F. (2018). Interactions between ecology and environmental science. *Environmental Science and Technology, 52*(14), 7632–7641.

Bloch, M. (1987). *The historian's craft*. Vintage Books.

Blumer, H. (1969). *Symbolic interactionism: Perspective and method*. University of California Press.

Brown, O., & Nicolucci-Altman, G. (2022). *The white paper on the future of environmental peacebuilding: Nurturing an ecosystem for peace* [White paper]. Geneva Peacebuilding Platform, International Union for Conservation of Nature, PeaceNexus Foundation, Environmental Law Institute, & Environmental Peacebuilding Association.

Bullard, R. (2000). *Dumping in Dixie: Race, class, and environmental quality*. Westview Press.

Carr, E. (1961). *What is history?* Random House.

Castells, M. (2012). *Networks of outrage and hope: Social movements in the Internet age*. Polity Press.

Conca, K., & Dabelko, G. (2002). *Environmental peacemaking*. Johns Hopkins University Press.

Conca, K., & Wallace, J. (2009). Environment and peacebuilding in war-torn societies: Lessons from the UN Environment Programme's experience with post-conflict assessment. *Global Governance, 15*(4), 485–504.

Darwin, C. (1859). *On the origin of species*. John Murray.

Dresse, A., Fischhendler, I., Nielsen, J., & Zikos, D. (2019). Environmental peacebuilding: Towards a theoretical framework. *Cooperation and Conflict, 54*(1), 99–119.

Dresse, C., Ide, T., & Ide, R. (2018). Environmental peacebuilding: A review of the empirical evidence and future research directions. *Journal of Peace Research, 55*(4), 491–503.

Durkheim, E. (1997). *The division of labor in society* (W. D. Halls, Trans.). Free Press. (Original work published 1893).

EcoPeace/Friends of the Earth Middle East. (2008). *Environmental peacebuilding theory and practice: A case study of the Good Water Neighbors Project and in-depth analysis of the Wadi Fukin/Tzur Hadassah communities*. https://ecopeaceme.org/wp-content/uploads/2022/03/environmental-peacebuilding-theory-and-practice-a-case-study-of-the-good-water-neighbours-projectand-in-depth-analysis-of-the-wadi-fukin-tzur-hadassah-communities.pdf

Ember, C., & Ember, M. (2014). *Cultural anthropology*. Pearson.

Escobar, A. (1999). *Development and its critics: The idea of progress in modern times*. Pluto Press.

Geertz, C. (1973). *The interpretation of cultures*. Basic Books.

Giddens, A. (2009). *Sociology* (6th ed.). Polity Press.

Heffington, E., Krampe, C., & Scheffran, J. (2019a). Environmental peacebuilding: A critical review of the literature. *Journal of Peace Research, 56*(5), 679–695.

Heffington, S., Oler, A., & Tretler, D. (Eds.). (2019b). *A national security strategy primer*. National Defense University Press.

Homer-Dixon, T. (1999). *Environment, scarcity, and violence*. Princeton University Press.

Ide, T. (2020, January 9). *Beware the dark side of environmental peacebuilding*. Climate Diplomacy. https://climate-diplomacy.org/magazine/conflict/beware-dark-side-environmental-peacebuilding

Ingold, T. (2000). *The perception of the environment: Essays on livelihood, dwelling, and skill*. Routledge.

Kottak, C. (2013). *Anthropology: The exploration of human diversity* (11th ed.). McGraw-Hill Education.

Koubi, V., Spilker, G., Böhmelt, T., & Bernauer, T. (2014). Do natural resources matter for interstate and intrastate armed conflict? *Journal of Peace Research, 51*(2), 227–243.

Krampe, F., & Swain, A. (2024). Environmental peacebuilding and conflict prevention: New insights and approaches. *International Environmental Agreements, 24*(1), 123–145.

Krampe, F., O'Driscoll, D., Johnson, M., Simangan, D., Hegazi, F., & de Coning, C. (2024). Climate change and peacebuilding: Sub-themes of an emerging research agenda. *International Affairs, 100*(3), 1111–1130.

Krugman, P., & Wells, R. (2018). *Economics* (5th ed.). Worth Publishers.

Lévi-Strauss, C. (1963). *Structural anthropology*. Basic Books.

Lunstrum, E. (2010). Reconstructing history, grounding claims to space: History, memory, and displacement in the Great Limpopo Transfrontier Park. *South African Geographical Journal, 92*(2), 129–143.

Mankiw, N. (2021). *Principles of economics* (9th ed.). Cengage Learning.

Marshall, A. (1890). *Principles of economics*. Macmillan and.

Marx, K. (1990). *Capital: A critique of political economy* (S. Moore, Trans.). Penguin Classics. (Original work published 1867).

Matthew, R., Brown, O., & Jensen, D. (2009). *From conflict to peacebuilding: The role of natural resources and the environment*. United Nations Environment Programme.

Matthew, R., Halle, M., & Switzer, J. (2002). *Conserving the peace: Resources, livelihoods, and security*. International Institute for Sustainable Development.

Millennium Ecosystem Assessment. (2005). *Ecosystems and human well-being: Synthesis*. Island Press.

Ostrom, E. (2009). A general framework for analyzing sustainability of social-ecological systems. *Science, 325*(5939), 419–422.

Parsons, T. (1951). *The social system*. Routledge.

Pellini, A., & Serrat, O. (2017). Enriching policy with research. In *Knowledge solutions: Tools, methods, and approaches to drive organizational performance* (pp. 937–944). Springer.

Pellow, D. N. (2000). *Resisting global toxics: Transnational movements for environmental justice*. MIT Press.

Ritzer, G. (2011). *Sociology: A global perspective* (8th ed.). McGraw-Hill Education.

Samuelson, P., & Nordhaus, W. (1992). *Economics: An introductory analysis* (14th ed.). McGraw-Hill.

Scoones, I. (1998). *Sustainable rural livelihoods: A framework for analysis*. IDS Working Paper 72.

Seixas, P., & Morton, O. (2013). *The big six: Historical thinking concepts*. The Historical Thinking Project.

Serrat, O. (2017). The sustainable livelihoods approach. In *Knowledge solutions: Tools, methods, and approaches to drive organizational performance* (pp. 21–26). Springer.

Serrat, O. (2025a). Environmental peacebuilding: Nurturing hope with analytical persuasion. In *Anthropogenic solutions for climate change: Achieving environmental peace* (pp. 165–170). Springer.

Serrat, O. (2025b). Harnessing evaluative criteria for environmental peacebuilding. In *Anthropogenic solutions for climate change: Achieving environmental peace* (pp. 171–208). Springer.
Serrat, O. (2025c). Disciplinary perspectives on environmental peacebuilding: Insights from environmental science and international relations. In *Anthropogenic solutions for climate change: Achieving environmental peace* (pp. 147–164). Springer.
Serrat, O. (2025d). Safeguarding Lake Chad. In *Anthropogenic solutions for climate change: Achieving environmental peace* (pp. 243–257). Springer.
Serrat, O. (2025e). Safeguarding the mother of waters: Environmental peacebuilding in the Mekong River Basin. In *Anthropogenic solutions for climate change: Achieving environmental peace* (pp. 229–239). Springer.
Slingerland, E. (2008). *What science offers the humanities: Integrating body and culture*. Cambridge University Press.
Smith, A. (1776). *The wealth of nations*. Methuen &, Ltd..
Sommer, U., & Fassbender, F. (2024). Environmental peacebuilding: Moving beyond resolving violence-ridden conflicts to sustaining peace. *World Development, 139*, 105348.
Thaler, R. (2015). *Misbehaving: The making of behavioral economics*. W. W. Norton & Company.
Tosh, J. (2015). *The pursuit of history: Aims, methods and new directions in the study of modern history* (6th ed.). Routledge.
United Nations Environment Programme. (2023). *Environmental cooperation and peacebuilding: A guide for practitioners*. United Nations Environment Programme.
United Nations Environment Programme. (2024, December 5). *Nature-based solutions for peace: Emerging practice and options for policymakers*. https://www.unep.org/resources/report/nature-based-solutions-peace-emerging-practice-and-options-policymakers
United Nations Environment Programme. (2025). *Environmental cooperation and peacebuilding*. https://www.unep.org/topics/disasters-and-conflicts/environment-security/environmental-cooperation-and-peacebuilding
Whorf, B. (1956). *Language, thought, and reality: Selected writings of Benjamin Lee Whorf*. MIT Press.

Chapter 4
Accounting for Complexity: Liberal Studies and Environmental Peacebuilding

Abstract Liberal studies, a multidisciplinary academic field spanning the arts, humanities, natural sciences, and social sciences, offers underutilized tools for reconciling competing ideologies, practices, and systems. This précis highlights how pluralist approaches enrich conceptual and evaluative frameworks, with special attention to context-specific criteria for assessing initiatives and generating lessons learned in environmental peacebuilding. Ultimately, liberal studies illuminates the theoretical foundations and practice of environmental peacebuilding and expands its capacity to foster durable, ecologically sound, and culturally attuned peace.

4.1 Re: Interdisciplinary Synthesis

Rooted in the liberal arts tradition, liberal studies serves as a crossroads of diverse ways of knowing—where the creative imagination of the arts, the interpretive subtlety of the humanities, the empirical rigor of the sciences, and the systemic insight of the social sciences converge. Anchored in a historical commitment to pluralism, the field cultivates the capacity to work productively across competing disciplinary frameworks. This orientation yields integrative insights and practical solutions, and this précis argues that liberal studies, through its commitment to pluralism, reflexivity, and methodological integration, equips environmental peacebuilding with the intellectual and practical tools to navigate complexity.

The intellectual framework of liberal studies clarifies how it catalyzes interdisciplinary research. Rooted in the humanistic and scientific traditions of the liberal arts, the field distinguishes itself through a deliberate commitment to integration as both method and mission. It does not merely juxtapose perspectives but actively fosters synthesis. This posture—historically attuned, philosophically grounded, and pragmatically responsive—treats complexity not as a problem to be solved but as a condition to be navigated (Serrat, 2017a). What does it mean to treat complexity as a condition to be navigated rather than a problem to be solved? This rhetorical shift underscores the distinctive ethos of liberal studies.

O. Serrat, *Advancing Environmental Peacebuilding*, SpringerBriefs in Political Science, https://doi.org/10.1007/978-981-92-4028-9_4

4.2 Historical and Philosophical Foundations of Liberal Studies

From its classical antecedents in the trivium and quadrivium—where grammar, logic, mathematics, and rhetoric formed a holistic intellectual training—liberal education has always sought to cultivate the capacity to think across domains and to connect knowledge to civic purpose (Kimball, 1986). Liberal studies inherits this ambition but reframes it for a contemporary scholarly landscape defined by complexity, uncertainty, and interdependence. It aims to nurture practitioners who can translate interdisciplinary insight into action.

Historically, the liberal arts tradition upheld the belief that knowledge should advance both personal growth and the common good. Liberal studies continues this dual commitment, reimagining it for the demands of contemporary research. Today's global challenges—climate change, conflict, economic precarity, entrenched social inequities, migration and displacement, public health crises, resource degradation, and technological disruption—defy resolution within disciplinary silos. Liberal studies responds with philosophical commitments to civic responsibility, dialogical engagement, and epistemic humility. These principles shape how questions are framed and methods selected.

4.3 Pluralism in Practice

Pluralism in liberal studies entails integrating diverse knowledge systems. However, meaningful incorporation requires more than token consultation; it requires genuine co-creation and methodological openness that treats local and scientific perspectives as coequal sources of insight. Practically, this involves co-designed research questions, participatory monitoring frameworks, and evaluation criteria that weave cultural, political, and relational indicators together with ecological and socioeconomic metrics. Without such pluralism, integration risks becoming tokenistic, reducing diverse knowledge systems to decorative add-ons rather than genuine sources of insight.

Liberal studies supplies the intellectual foundation for this work by treating multiplicity as philosophy and practice. It encourages fluid movement across scales of analysis—from local practices to regional and international policy—and across forms of evidence—from statistical models to oral histories. Eschewing a patchwork of perspectives, liberal studies cultivates a dialogic, iterative process through which complementary methods refine one another to produce more ecologically informed, equitable, and robust solutions.

4.4 Methodological Pathways for Integration

If the philosophical commitments of liberal studies provide the "why" of integrative scholarship, its interdisciplinary methodologies supply the "how." At the heart of liberal studies lies methodological diversity, treating disciplinary boundaries not as barriers but as porous lines through which ideas, perspectives, and practices circulate. This approach begins with a recognition that each discipline offers partial but valuable insights into complex phenomena.

One common strategy is the interdisciplinary research process outlined by Repko et al. (2019), which proceeds from problem definition to disciplinary insight gathering, integration, and understanding. In practice, this might mean combining the statistical modeling of environmental science with the ethnographic depth of anthropology, the policy analysis of political science, and the ethical reasoning of philosophy. Such integration is not without friction: quantitative and qualitative paradigms often clash in their assumptions about causality, generalizability, and validity. Liberal studies addresses these tensions through methodological transparency and reflexive meta-frameworks that articulate epistemological commitments.

4.5 Managing Epistemic Tensions

A defining challenge of interdisciplinary work is reconciling divergent methodological traditions. Quantitative approaches prioritize generalizability and replicability, while qualitative approaches emphasize contextual depth and interpretive nuance. Liberal studies confronts these tensions not by privileging one paradigm but by designing frameworks in which methods test and enrich one another.

By embedding synthesis into design, liberal studies transforms methodological diversity into a source of rigor: each method is tested and refined against others. Critics contend that pluralism risks diluting disciplinary rigor, leading to incoherence or superficiality. Yet liberal studies reframes rigor as the ability to integrate across paradigms, not merely to enforce standards within one, and it does so by insisting on reflexivity, transparency, and shared standards that ensure synthesis yields integrated explanations rather than loose assemblages.

4.6 Environmental Peacebuilding as an Interdisciplinary Challenge

The need for the integrative ethos of liberal studies is particularly salient in environmental peacebuilding, a field that explores how environmental cooperation can help prevent conflict and foster lasting peace. Environmental peacebuilding is inherently interdisciplinary (Serrat, 2025a, 2025b). Addressing its challenges requires ecological science to assess conditions and risks, political analysis to examine governance and conflict dynamics, and social inquiry to illuminate community relationships and cultural narratives. No single discipline can capture the feedback loops linking environmental degradation, resource competition, and social (in)stability. Instead, integrated evidence informs policy instruments such as community-based resource management and transboundary agreements, enabling responses that are both context-sensitive and durable.

The interdisciplinary requirements of environmental peacebuilding have deepened as the field itself has evolved. Early debates framed the environment primarily through the lens of environmental security, but scholarship has since shifted toward recognizing environmental cooperation as a pathway to peace (Conca & Dabelko, 2002; Ide et al., 2021). This evolution underscores the methodological openness that aligns with liberal studies, combining empirical data with normative analysis and integrating local knowledge with global policy.

4.7 Theoretical Lenses for Pluralist Inquiry

Theoretical lenses provide conceptual frameworks to frame, interpret, and evaluate environmental peacebuilding research. In the liberal studies tradition, these lenses are complementary tools that illuminate different facets of complex problems. For example, constructivism highlights how ideas, norms, and identities shape cooperation; environmentalism centers ecological integrity and sustainability; feminism emphasizes gendered governance and inclusive participation; postcolonialism interrogates historical inequalities and power asymmetries; systems theory emphasizes interconnected subsystems and adaptive interventions. Other lenses, such as critical theory, Indigenous epistemologies, political economy, or pragmatism, further enrich this pluralist repertoire. Each lens suggests distinct evaluative priorities, and together they inform pluralist criteria that are richer than any single framework could provide. No single lens suffices; together, these perspectives generate evaluative standards that are more comprehensive than any one approach.

4.8 Illustrative Cases in Practice

Far from being automatic, interdisciplinarity in environmental peacebuilding must be intentionally forged—a reality that case studies bring to light as they reveal how research designs adapt across different socio-ecological and political settings:

- **Senegal River Basin Development Organization.** Founded on the principle of "common and indivisible property," this initiative drew on anthropology, hydrological science, and political science (Bolognesi et al., 2015). The integration of local fishing communities' knowledge into basin management ensured solutions were socially legitimate and culturally resonant. Evaluative criteria extended beyond distributional efficiency to encompass conflict prevention, equitable benefit sharing, and ecological sustainability. This case illustrates how pluralist evaluation criteria can transform technical cooperation into socially legitimate governance.

- **UNEP's *Greening the Blue Helmets* Initiative.** This initiative integrated environmental management into peacekeeping operations, drawing on and conflict resolution, ecology, and logistics (United Nations Environment Programme, 2012). Evaluations considered not only operational efficiency and environmental impact but also contributions to trust building and local capacity. Here, pluralism enabled the synthesis of diverse disciplinary inputs into a coherent strategy, reflexively adapted through feedback from peacekeepers and local stakeholders.
- **Great Limpopo Transfrontier Park.** This initiative combined conservation biology, postcolonial critique, and tourism economics to address biodiversity protection and the socio-economic needs of communities historically displaced by conservation policies (Hoogendoorn et al., 2019). Evaluative criteria incorporated ecological indicators, income generation metrics, and participatory governance. This case demonstrates pluralistic commitment to balancing ecological, economic, and social dimensions.

Together, these cases show that interdisciplinarity is not an abstract ideal but a pragmatic necessity for durable and just interventions.

4.9 Pluralist Metrics for Environmental Peacebuilding

Environmental peacebuilding gains practical force when it rests on robust evaluation systems (Serrat, 2025c). In the liberal studies tradition, evaluation is reflective, drawing on cultural perspectives, disciplinary standards, and ethical commitments. If peacebuilding is judged only by reductions in violence, it risks overlooking whether peace is just, inclusive, and ecologically sustainable.

In international development, evaluation frameworks draw on the Organisation for Economic Co-operation and Development (OECD) Development Assistance Committee's criteria—relevance, coherence, effectiveness, efficiency, impact, and sustainability (OECD, 2019; Serrat, 2017b). Liberal studies would enrich these by, for example, expanding "impact" to include ecological restoration and empowerment, and by redefining "efficiency" to recognize that slower, participatory processes may yield more durable peace.

Building on this foundation, new evaluative criteria tailored to fragile, resource-dependent contexts might include adaptability and resilience, conflict sensitivity and "do no harm", cultural sensitivity, inclusivity and participation, legitimacy, long-term environmental impact, and "peace writ large" (Serrat, 2025d). By establishing clear standards that integrate the hybrid cultural, ecological, economic, normative, political, security, and social contexts of peacebuilding, such criteria would help identify and promote replicable good practices across diverse settings. They would guide the design, implementation, and assessment of initiatives, while enhancing their comparability, long-term impact, and scalability. In this way, evaluation would be positioned as a driver of innovation and resilience, instead of being relegated to an end-of-project obligation. Critics might argue pluralist evaluative criteria risk being too diffuse or difficult to measure. Yet liberal studies reframes evaluation as adaptive and transparent, where rigor comes from contextual fit rather than universal metrics. Taken together, such new evaluative criteria would demonstrate that pluralism is not only a methodological choice but also a strategic imperative. They point toward broader lessons for future research and practice, where liberal studies can shape both scholarly inquiry and policy design.

4.10 Lessons for Future Research and Practice

This précis has shown in particular how liberal studies provides a valuable framework for advancing environmental peacebuilding. In this evolving field, future research must not only refine integrative methodologies that respond to shifting environmental and geopolitical realities but also embed pluralist evaluation criteria into the design of initiatives. In doing so, pluralism becomes institutionalized rather than episodic, shaping policy, funding, and implementation at scale (Serrat, 2025e, 2025f). While skeptics might contend that pluralist frameworks are too ambitious to operationalize broadly, case studies suggest otherwise: when embedded from the outset, pluralism enhances legitimacy, adaptability, and long-term resilience. The question is no longer whether environmental peacebuilding can afford pluralism, but whether it can endure without it.

References

Bolognesi, T., Bréthaut, C., Sangbana, K., & Tignino, M. (2015). *Transboundary governance in the Senegal and Niger River Basins: Historical analysis and overview of benefit-sharing arrangements*. Geneva Water Hub. https://www.genevawaterhub.org/sites/default/files/atoms/files/gwh_note_on_water_governance_in_omvs_and_abn_final.pdf

Conca, K., & Dabelko, G. (Eds.). (2002). *Environmental peacemaking*. Johns Hopkins University Press.

Hoogendoorn, G., Kelso, C., & Sinthumule, I. (2019). Tourism in the Great Limpopo Transfrontier Park: A review. *African Journal of Hospitality, Tourism and Leisure, 8*(5).

Ide, T., Krampe, F., & Swain, A. (2021). Environmental peacebuilding: Towards a theoretical framework. *Cooperation and Conflict, 56*(1), 3–26.

Kimball, B. (1986). *Orators & philosophers: A history of the idea of liberal education*. College Board.

Organisation for Economic Co-operation and Development. (2019). *Evaluation criteria: Adapted definitions and principles for use*. https://one.oecd.org/document/DCD/DAC(2019)58/FINAL/en/pdf

Repko, A., Szostak, R., & Buchberger, M. (2019). *Introduction to interdisciplinary studies* (3rd ed.). SAGE.

Serrat, O. (2017a). Understanding complexity. In *Knowledge solutions: Tools, methods, and approaches to drive organizational performance* (pp. 345–353). Springer.

Serrat, O. (2017b). Learning from evaluation. In *Knowledge solutions: Tools, methods, and approaches to drive organizational performance* (pp. 879–884). Springer.

Serrat, O. (2025a). Disciplinary perspectives on environmental peacebuilding: Insights from environmental science and international relations. In *Anthropogenic solutions for climate change: Achieving environmental peace* (pp. 147–164). Springer.

Serrat, O. (2025b). *Bridging the gap: Synergizing anthropology, economics, history, and sociology for environmental peacebuilding*. Unpublished manuscript, Georgetown University. https://www.researchgate.net/publication/391522222_Bridging_the_Gap_Synergizing_Anthropology_Economics_History_and_Sociology_for_Environmental_Peacebuilding

Serrat, O. (2025c). Environmental peacebuilding: Nurturing hope with analytical persuasion. In *Anthropogenic solutions for climate change: Achieving environmental peace* (pp. 165–170). Springer.

Serrat, O. (2025d). Harnessing evaluative criteria for environmental peacebuilding. In *Anthropogenic solutions for climate change: Achieving environmental peace* (pp. 171–208). Springer.

Serrat, O. (2025e). *Milestones of scholarship: From idea to contribution in environmental peacebuilding evaluation*. Unpublished manuscript, Georgetown University. https://www.researchgate.net/publication/397223335_Milestones_of_Scholarship_From_Idea_to_Contribution_in_Environmental_Peacebuilding

Serrat, O. (2025f). *Assessing environmental peacebuilding: Principles, criteria, and pathways for proactive interventions*. Unpublished manuscript, Georgetown University. https://www.researchgate.net/publication/398543700_Assessing_Environmental_Peacebuilding_Principles_Criteria_and_Pathways_for_Proactive_Interventions

United Nations Environment Programme. (2012). *Greening the Blue Helmets: Environment, natural resources and UN peacekeeping operations*. UNEP. https://wedocs.unep.org/handle/20.500.11822/8840

Chapter 5
Milestones of Scholarship: From Idea to Contribution in Environmental Peacebuilding

Abstract This précis traces the scholarly process as a progression from idea to contribution and illustrates it through what would be an exemplar thesis in the emerging field of environmental peacebuilding. The first part outlines how ideas are disciplined into concepts, hypotheses, arguments, and contributions, each marked by milestones that transform intuition into knowledge. The second part applies this framework to the development of context-specific evaluative criteria in environmental peacebuilding, highlighting the limitations of standardized frameworks and the need for tailored approaches. Together, the précis demonstrates how rigorous scholarship can generate both theoretical advances and practical tools for more effective, context-specific interventions.

5.1 Introduction

The defining hallmark of scholarship is its iterative progression through key milestones—idea, conceptualization, hypothesis, argument, and contribution—each revisited and reshaped through feedback, critique, and new evidence, shaping the intellectual trajectory and practical relevance of inquiry (Booth et al., 2016). While the initial spark may arise from exigency, intuition, or observation, the scholarly process disciplines these impulses into concepts, refines them into hypotheses, and elaborates them into arguments capable of withstanding scrutiny. This transformation is neither linear nor mechanical; it unfolds through spiraling, dialogic cycles of exchange and refinement, in which earlier stages are continually re-examined and provisional claims revised, building cumulatively on prior insights, drawing on established traditions of inquiry while seeking to extend or challenge them. Scholarship thus offers not only a methodological pathway but also a set of evaluative milestones that function less as fixed endpoints than as generative nodes, seeding subsequent cycles through which knowledge claims become credible, communicable, and impactful (Booth et al., 2016).

This précis argues that the scholarly process offers a disciplined pathway for transforming intuition into knowledge, illustrated through an exemplar thesis on

O. Serrat, *Advancing Environmental Peacebuilding*, SpringerBriefs in Political Science, https://doi.org/10.1007/978-981-92-4028-9_5

context-specific evaluative criteria in the emerging field of environmental peacebuilding. Environmental peacebuilding refers to the integration of natural resource management and environmental cooperation into strategies for conflict prevention, peacebuilding, and post-conflict recovery (Conca & Dabelko, 2002; Ide, 2019; Swain & Öjendal, 2018).

This précis advances two interwoven aims. First, it charts the formal progression of scholarship, clarifying stage-specific deliverables and emphasizing the dynamic interplay among conceptual clarity, methodological rigor, and evidentiary grounding. Second, it animates this arc through a novel and timely case in environmental peacebuilding: a thesis that would develop context-specific evaluative criteria. The thesis would identify a conceptual gap, viz., the inadequacy of universalized evaluation frameworks, would formulate a testable hypothesis, and would build an argument with both theoretical and practical significance. Crucially, each of these stages is provisional: the identification of the gap would be sharpened by feedback from the literature, the hypothesis would be revised in light of empirical encounters, and the argument would be re-anchored through dialogue with both scholarly and practitioner audiences. By juxtaposing scholarly architecture with concrete application, this précis demonstrates that intellectual rigor and applied relevance are not only compatible but mutually reinforcing.

5.2 Milestones of Scholarship: From Idea to Contribution

The progression from idea to contribution is best understood as an iterative sequence of intellectual refinements, more like a spiral than a straight road, where each stage may be revisited and revised, marked by distinct methodological practices and milestones (Booth et al., 2016). Appendix A specifies five milestones, noting their respective pathways and dimensions of practice.

5.2.1 Disciplining Intuition: From Idea to Conceptualization

Every scholarly endeavor begins with an idea—but in its raw form, an idea is insufficient for systematic inquiry. Ideas are often intuitive, impressionistic, or sparked by immediate observation: a recognition of a gap in existing knowledge, a contradiction in the literature, or a pressing problem in practice. At this stage, the idea is generative but unstable; it gestures toward a direction of thought without yet providing the scaffolding required for rigorous investigation. The scholar's first task, therefore, is to discipline the idea into a concept.

Conceptualization clarifies, delimits, and communicates an idea by identifying terms of debate, defining key constructs, and situating the inquiry within existing scholarship. Yet conceptualization is rarely settled in a single pass: definitions and constructs are provisional, often re-framed as critiques, empirical puzzles, or new literature emerge. This stage is not merely preparatory but constitutive: how a concept is framed shapes the research question (and subquestions) posed and the evidence deemed valid. For example, defining "environmental peacebuilding" as institutional capacity-building leads to a different trajectory than framing it as the cultivation of ecological resilience or trust recovery (Conca & Dabelko, 2002; Ide, 2019; Swain & Öjendal, 2018). Conceptualization therefore requires both analytical precision and interpretive judgment. Its deliverable is the literature review, which maps prior scholarship to establish the inquiry's novelty and necessity while anchoring the emerging concept in a recognizable scholarly conversation. In this process, private intuition becomes a public claim—open to contestation, evaluation, and refinement—marking the scholar's entry into the collective enterprise of knowledge production.

5.2.2 Framing Inquiry: From Conceptualization to Hypothesis

Once an idea has been disciplined into a concept and situated within scholarly conversation, the next step is to translate conceptual clarity into a hypothesis. Whereas conceptualization defines the terms of inquiry, hypothesis formation articulates the claims that can be debated, tested, or defended. A hypothesis is not a prediction; it is a structured proposition that links concepts to observable or arguable outcomes. In this sense, it serves as a bridge between the abstract realm of theory and the empirical or analytical work that follows.

Hypothesis formation requires operationalization: the careful translation of abstract concepts into measurable indicators. For instance, if a scholar conceptualizes "trust" as a central dimension of environmental peacebuilding, the hypothesis must specify how trust will be identified, measured, or inferred—whether through institutional arrangements, patterns of cooperation, or survey data. Because hypotheses are provisional by design, they are often reformulated in response to peer critique, feasibility constraints, or pilot evidence, underscoring the iterative movement between conceptualization and hypothesis.

The deliverable at this stage is the research design: a blueprint that articulates the hypothesis, outlines the methodology, and identifies sources of evidence. The research design serves both as a planning tool and a test of scholarly rigor, persuading peers that the hypothesis is significant and researchable. Crucially, it marks a shift in accountability—from the inward-facing clarity of conceptualization to the outward-facing demands of critique, evaluation, and refinement.

To advance, the inquiry must shift from conceptual clarity to structured testing. This marks a deepening of scholarly commitment: no longer mere curiosity, the endeavor now carries defined stakes, methods, and anticipated contributions. It lays the groundwork for argument development in general, where hypotheses are tested, evidence assembled, and claims shaped into a coherent, persuasive whole.

5.2.3 Building Claims: From Hypothesis to Argument

The shift from hypothesis to argument marks scholarship's move from possibility to proof. However well-crafted, a hypothesis remains provisional until tested against evidence and developed into sustained reasoning. At this stage, the scholar must both interpret evidence relevant to the hypothesis and shape it into a coherent, defensible argument.

Evidence gathering may take diverse forms—archival research, comparative case studies, expert interviews, fieldwork, or statistical analysis—depending on disciplinary norms and the nature of the inquiry. Yet its purpose remains consistent: to provide a foundation upon which claims can be evaluated. Evidence only gains meaning through interpretation. It must be deliberately chosen, framed within a study's concepts, and analyzed in light of the methods guiding the research. This interpretive act is what transforms data into insight.

Constructing an argument involves more than assembling supportive evidence. It requires synthesis: the integration of disparate findings into a structured narrative that demonstrates that the hypothesis is plausible—and why it matters. A strong argument situates its claims within broader scholarly debates, anticipates counterarguments, and bears out the robustness of its reasoning. In this sense, argumentation is both defensive and generative—defending the validity of the hypothesis while generating new perspectives that extend beyond the initial proposition.

The deliverable at this stage is the production of concept papers, conference presentations, draft chapters, policy briefs, or working papers. These interim outputs serve as testing grounds, enabling the scholar to refine the argument through dialogue with peers. Feedback reveals weaknesses in logic, gaps in evidence, or ambiguities that might compromise the final contribution. Often, this feedback compels the scholar to loop back—revising hypotheses, re-defining concepts, or even reframing the initial idea—illustrating the spiral nature of scholarly development.

Ultimately, the movement from hypothesis to argument represents the maturation of scholarly inquiry. What began as an idea, disciplined into a concept and specified as a hypothesis, now emerges as a fully articulated argument capable of advancing knowledge. This stage underscores the cumulative and dialogic nature of scholarship: arguments are not final pronouncements but evolving positions, shaped by critique and open to refinement as new evidence and perspectives emerge.

5.2.4 Making a Difference: From Argument to Contribution

The culmination of the scholarly process lies not just in the construction of an argument, but in its transformation into a recognized contribution to the body of knowledge. However well-reasoned, an argument achieves significance only when placed within broader intellectual and practical conversations. This involves refining the argument for clarity and rigor, subjecting it to critical evaluation, and situating it within existing debates and applications.

Refinement comes first. Draft arguments are rarely definitive; they demand revision through peer review, supervisory feedback, or self-assessment. Through repeated critique and revision, arguments gain clarity, precision, and persuasiveness. Far from mechanical, refinement is a creative act of scholarly craftsmanship that makes the argument precise and compelling.

The second task is evaluation through scholarly dialogue. Peer review, conference presentations, and workshops serve as crucibles where arguments are tested under scrutiny. This stage underscores the communal nature of scholarship: knowledge advances not in isolation, but through collective interrogation and refinement. The willingness to expose one's work to critique is not ancillary, but essential to producing credible scholarship.

Finally, contribution means situating the argument within its field. Whether advancing theory, innovating method, or offering policy insight, each contribution extends knowledge or reframes problems. The milestone is the final manuscript—article, dissertation, or monograph—that distills the argument into durable form. From here, the work enters the archive of collective knowledge, ready to be built upon, contested, or adapted. In this sense, contributions are not endpoints but generative nodes, sparking new questions and inaugurating subsequent cycles of inquiry.

The progression from argument to contribution completes the scholarly cycle: a process that moves from initial insight to refined contribution. Yet the process is neither linear nor final—each contribution sparks further questions, renewing the cycle. Scholarship thus emerges as a dynamic practice—at once creative in its individuality and sustained through collective dialogue.

5.3 Exemplar in Practice: Evaluating Environmental Peacebuilding

5.3.1 *Identifying the Gap: Idea*

The scholarship process outlined above could be illustrated by the development of a thesis on context-specific evaluative criteria in environmental peacebuilding. The initial idea would arise from a practical observation: despite the proliferation of peacebuilding initiatives that integrate environmental dimensions—ranging from transboundary water management to post-conflict natural resource governance—evaluation frameworks remain predominantly standardized. In international development, the most commonly employed framework is drawn from the Organisation for Economic Co-operation and Development's Development Assistance Committee (OECD, 2019), whose six evaluation criteria are relevance (alignment with local needs), coherence (compatibility with other interventions and policy frameworks),[1] effectiveness (achievement of objectives), efficiency (optimal resource use), impact (long-term changes in conflict and environmental health), and sustainability (durability of outcomes).

Though widely applied by donors, governments, and implementing agencies in development, the OECD–DAC framework does not capture the nuanced dynamics—such as ecological interdependence, local legitimacy, and trust-building—that shape outcomes in fragile contexts (Ide, 2019; Swain & Öjendal, 2018). Environmental peacebuilding therefore requires evaluative criteria that account for complexity, leverage multidisciplinary perspectives, and are adapted to context (Serrat, 2025a, 2025b, 2025c, 2025d).

5.3.2 *Mapping the Literature: Conceptualization*

To move from idea to concept, the scholar would situate this intuition within existing debates. Conceptualization would involve mapping the literature on peacebuilding evaluation, environmental governance, and adaptive management, and identifying the relevant gap. While evaluation scholarship emphasizes rigor and comparability, and environmental peacebuilding literature foregrounds the importance of local dynamics, no work has integrated the two. The emerging concept would be a framework of context-specific evaluative criteria—a set of measures capable of adapting to local conditions while maintaining analytical coherence.

The deliverable at this stage would be the literature review, which establishes both the novelty and necessity of the inquiry. By showing that existing frameworks inadequately capture context, and by drawing on adjacent literatures that stress adaptability and inclusivity, the scholar would transform the initial idea into a concept that is both researchable and relevant. This conceptualization would set the stage for the next step: the formulation of a hypothesis about the value and impact of context-specific evaluative criteria in practice.

5.3.3 *Formulating the Proposition: Hypothesis*

Having established context-specific evaluative criteria as a necessary complement to standardized frameworks, the next step would be to translate this conceptual clarity into a testable hypothesis. The central proposition would be that frameworks tailored to local ecological, social, and political contexts yield more accurate, legitimate, and actionable assessments of environmental peacebuilding initiatives than universalized evaluation criteria. This hypothesis is not merely descriptive; it is

[1] Introduced in 2019, "coherence" serves to avoid contradictions across agendas.

normative, empirically tractable, and invites systematic inquiry into whether—and how—context-specific approaches outperform generic ones.

Operationalization is essential at this stage. The abstract notion of "context-specific criteria" must be translated into identifiable dimensions that can be observed and assessed. Serrat (2025e) surmises that such evaluative criteria for environmental peacebuilding initiatives might be organized across community, conflict, cultural, economic, environmental, and political dimensions. Together, these dimensions provide a structured foundation for comparative analysis, ensuring that the inquiry remains both theoretically coherent and methodologically robust.

The deliverable at this stage would be the research design (or proposal): it would define the hypothesis, outline the methodology, and identify sources of evidence. A comparative case study approach would provide a strong strategy for tracing the diverse pathways and constraints of environmental peacebuilding (Swain & Öjendal, 2018). To capture variation across scales and settings, potential cases would span different ecological systems, governance levels, and geopolitical contexts: Lake Chad, which underscores the urgency of resource-sharing under devastating ecological stress; the Heart of Borneo Initiative, which models conservation-led cross-border cooperation at the national and regional level; the Mekong River, where transboundary governance entangles geopolitical tensions with local livelihoods; the Amazon, which reveals the nexus of ecological preservation, indigenous rights, and global climate concerns; and the Indus Waters Treaty, which illustrates the resilience of institutionalized cooperation amid protracted conflict (Serrat, 2025f, 2025g, 2025h). Together, these cases would demonstrate the breadth of environmental peacebuilding challenges—spanning lakes and rivers, forests and watersheds, national initiatives and international treaties, and extending from regional to global coverage. Where feasible, site visits to these regions would provide invaluable opportunities for direct observation and engagement with local actors, grounding the analysis in lived realities. Yet such visits would also confront practical constraints—difficulties of access, security risks in conflict-affected zones, and the ethical responsibility to avoid burdening vulnerable communities.

The research design would serve both as a blueprint for inquiry and as a persuasive case for the hypothesis's significance, feasibility, and scholarly value. By moving from concept to hypothesis, the thesis would become a structured project with defined claims, methods, and anticipated contributions. This stage would lay the groundwork for argument development, where evidence would be gathered and synthesized to substantiate the relevance of context-specific evaluative criteria.

5.3.4 Assembling the Case: Argument

The transition from hypothesis to argument marks the point where provisional claims are tested against evidence and developed into sustained reasoning. At this juncture, the scholar's task is twofold: to gather and interpret evidence that directly addresses the hypothesis, and to weave those findings into a coherent argument capable of withstanding scrutiny.

Evidence gathering can take diverse forms—archival research, comparative case studies, expert interviews, fieldwork, or statistical analysis—depending on disciplinary norms and the nature of the inquiry. But evidence does not speak for itself: it must be selected, contextualized, and interpreted in light of the conceptual framework and methodological commitments established earlier. This interpretive act is what turns data into insight.

Constructing an argument requires more than assembling evidence—it demands synthesis: weaving disparate findings into a narrative that intimates plausibility and significance. Strong arguments engage broader debates, anticipate counterpoints, and display reasoning. For example, the OECD–DAC framework promotes comparability, accountability, and scalability—useful for benchmarking and resource allocation (OECD, 2019). Yet in fragile contexts, these strengths can become liabilities: comparability may obscure local nuance, accountability can privilege upward reporting over community legitimacy, and scalability assumes uniform conditions rarely present in fragile or conflict-affected settings. Context-specific evaluative criteria, by contrast, would preserve rigor while capturing dynamics, e.g., distributive equity, ecological interdependence, trust-building, that standardized models often miss. The exemplar would not reject standardization outright but would highlight its limits when applied uncritically to environmental peacebuilding. The deliverable at this stage would be interim outputs, specified earlier, that test the argument through peer dialogue. Feedback exposes gaps in logic, evidence, or framing, enabling refinement and sharpening the argument.

Ultimately, the movement from hypothesis to argument would represent the maturation of scholarly inquiry. What began as an idea, disciplined into a concept and specified as a hypothesis, would now emerges as a fully articulated argument capable of advancing knowledge. This stage would underscore the cumulative and dialogic nature of scholarship: arguments are not final pronouncements but contributions to an ongoing conversation, open to refinement and contestation as new evidence and perspectives emerge.

5.3.5 Distilling the Contribution: Advancing Evaluation in Context

At this juncture, the thesis would convert the argument—that context-specific evaluative criteria yield more accurate, legitimate assessments of environmental peacebuilding—into a recognized scholarly contribution, demonstrating its significance not only through evidence but also through relevance to academic debate and practical application.

Refinement would be the first step. The argument developed through case studies and comparative analysis would have to be revised in response to peer feedback, supervisory critique, and critical self-assessment. This could involve clarifying the scope of context-specificity, sharpening the logic linking evaluative criteria to peacebuilding outcomes, or addressing counterarguments that emphasize the value of standardized comparability. Such refinement would ensure the argument is both internally coherent and externally persuasive.

The second step would be evaluation through scholarly dialogue. Presenting findings at, say, conferences or in working papers would expose the argument to peer scrutiny. In environmental peacebuilding, this dialogue may need to engage both academic audiences and practitioner communities, whose insights would clarify the practical feasibility and relevance of context-specific frameworks. This stage would underscore the communal nature of scholarship: contributions gain credibility through engagement and contestation—not in isolation.

Lastly, contribution would mean positioning the argument within its field. Theoretically, the thesis would advance evaluation scholarship by embedding context-sensitivity into peacebuilding assessment and challenging universalized frameworks. Methodologically, it would determine how context-specific evaluative criteria can be operationalized without sacrificing rigor. Practically, it would equip policymakers and practitioners with adaptable tools for monitoring and evaluation in fragile settings. If feasible, site visits to case study regions would further strengthen this contribution, grounding the evaluative criteria in lived realities. The deliverable at this stage would be a manuscript that distills the argument into a durable contribution to knowledge and practice.

In this way, the exemplar would reflect the general scholarly process, moving from initial observation to a refined contribution. Thus, the thesis would illustrate how scholarship can simultaneously advance theoretical debates and generate tools for more effective and just interventions in environmental peacebuilding. The milestones, activities, and indicative timing for each stage of thesis development are presented in Appendix B.

5.4 Conclusion: Scholarship as a Vehicle for Contribution

Scholarship unfolds through successive, iterative milestones that transform private intuitions into public insights. By tracing this progression, this précis has shown that intellectual rigor and practical relevance are not competing but complementary commitments.

The exemplar would demonstrate the scholarly process in practice. It would highlight that while standardized frameworks such as the OECD–DAC provide essential benchmarks for comparability and accountability, they often miss the nuanced dynamics most critical in fragile contexts (Serrat, 2025i). Tailored evaluative criteria, by contrast, would surface dimensions—such as ecological resilience, local legitimacy, and trust-building—that standardized models may overlook (Ide, 2019). Acknowledging the OECD–DAC framework's widespread adoption and institutional embeddedness, the exemplar would also enrich it with context-sensitive adaptations, aligning the dominant model more closely with the lived realities and complex interdependencies of fragile environments.

The resulting contribution would be twofold: theoretically, it would advance evaluative scholarship by embedding context-sensitivity into environmental peacebuilding assessment; practically, it would equip policymakers and practitioners with adaptable tools for fragile environments. Taken together, the scholarly process and exemplar would explicate that academic inquiry is not merely a pathway from idea to contribution but a vehicle for producing contributions that matter—expanding knowledge while guiding effective interventions.

Appendix A

See Fig. 5.1

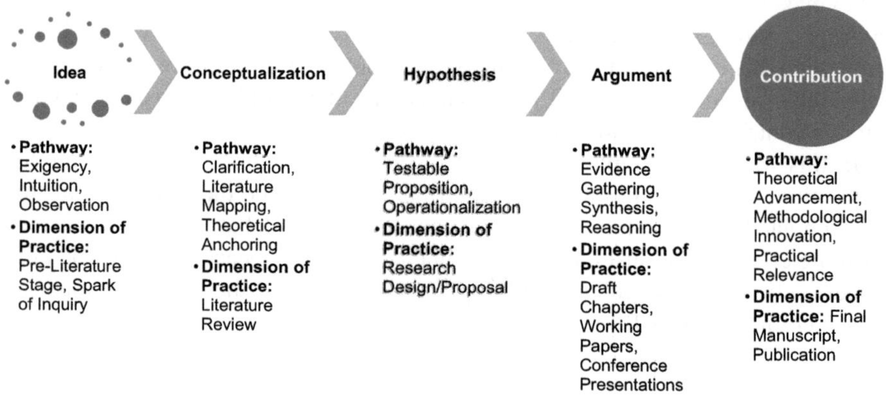

Fig. 5.1 Milestones of Scholarship

Appendix B

See Table 5.1

Table 5.1 Milestones and Timeline of Proposed Thesis

Milestone	Activity	Indicative Timing (Months)
Idea → Conceptualization	• Identify research problem and initial idea • Conduct literature review • Define key concepts and situate within debates • Draft research question (and subquestions)	Months 1–3
Conceptualization → Hypothesis	• Operationalize concepts into variables/criteria • Formulate hypothesis or guiding framework • Draft detailed research design/proposal • Secure ethics approval by the Institutional Review Board (if needed) • Defend proposal	Months 4–6
Hypothesis → Argument	• Conduct archival research • Conduct comparative case studies across multiple regions (e.g., Amazon, Borneo, Indus, Lake Chad, Mekong). • Gather and analyze data • Draft early chapters/working papers • Present preliminary findings at a conference/ seminar	Months 7–14
Argument → Contribution	• Complete data analysis and synthesis • Draft full thesis chapters • Make iterative revisions in response to peer/ supervisor feedback • Circulate near final draft	Months 15–20
Finalization & Dissemination	• Submit full draft for review • Revise full draft post review • Defend thesis (oral viva/defense) • Revise full-draft post-defense if required • Submit final manuscript • Prepare publications (articles, policy briefs)	Months 21–24

References

Booth, W., Colomb, G., & Williams, J. (2016). *The craft of research* (4th ed.). University of Chicago Press.

Conca, K., & Dabelko, G. (Eds.). (2002). *Environmental peacemaking*. Woodrow Wilson Center Press.

Ide, T. (2019). The impact of environmental cooperation on peacemaking: Definitions, mechanisms, and empirical evidence. *International Studies Review, 21*(3), 327–346. https://doi.org/10.1093/isr/viy014

OECD. (2019). *Better criteria for better evaluation: Revised evaluation criteria definitions and principles for use*. OECD Publishing. https://www.oecd.org/en/publications/better-criteria-for-better-evaluation_15a9c26b-en.html

Serrat, O. (2025a). Harnessing evaluative criteria for environmental peacebuilding. In *Anthropogenic solutions for climate change: Achieving environmental peace* (pp. 171–208). Springer.

Serrat, O. (2025b). *Accounting for complexity: Liberal studies and environmental peacebuilding*. Unpublished manuscript, Georgetown University. https://www.researchgate.net/publication/396824279_Accounting_for_Complexity_Liberal_Studies_and_Environmental_Peacebuilding

Serrat, O. (2025c). Disciplinary perspectives on environmental peacebuilding: Insights from environmental science and international relations. In *Anthropogenic solutions for climate change: Achieving environmental peace* (pp. 147–164). Springer.

Serrat, O. (2025d). *Bridging the gap: Synergizing anthropology, economics, history, and sociology for environmental peacebuilding*. Unpublished manuscript, Georgetown University. https://www.researchgate.net/publication/391522222_Bridging_the_Gap_Synergizing_Anthropology_Economics_History_and_Sociology_for_Environmental_Peacebuilding

Serrat, O. (2025e). Environmental peacebuilding: Nurturing hope with analytical persuasion. In *Anthropogenic solutions for climate change: Achieving environmental peace* (pp. 165–170). Springer.

Serrat, O. (2025f). Safeguarding Lake Chad. In *Anthropogenic solutions for climate change: Achieving environmental peace* (pp. 243–257). Springer.

Serrat, O. (2025g). Preserving paradise: The Heart of Borneo Initiative. In *Anthropogenic solutions for climate change: Achieving environmental peace* (pp. 221–227). Springer.

Serrat, O. (2025h). Safeguarding the mother of waters: Environmental peacebuilding in the Mekong River Basin. In *Anthropogenic solutions for climate change: Achieving environmental peace* (pp. 229–239). Springer.

Serrat, O. (2025i). *Assessing environmental peacebuilding: Principles, criteria, and pathways for proactive interventions*. Unpublished manuscript, Georgetown University. https://www.researchgate.net/publication/398543700_Assessing_Environmental_Peacebuilding_Principles_Criteria_and_Pathways_for_Proactive_Interventions

Swain, A., & Öjendal, J. (Eds.). (2018). *Routledge handbook of environmental conflict and peacebuilding*. Routledge.

Chapter 6
Assessing Environmental Peacebuilding: Principles, Criteria, and Pathways for Proactive Interventions

Abstract Drawing on comparative case studies, this précis contends that environmental peacebuilding can be promoted through three guiding principles: shared resource cooperation, sustainability in peace processes, and inclusive natural resource governance. However, principles alone are insufficient—practice must navigate the hybrid ecological, political, and social contexts of environmental peacebuilding. Accordingly, this précis proposes adaptations to the OECD–DAC framework and supplements it with peacebuilding-specific evaluative criteria, enabling more holistic design, implementation, and assessment of interventions.

6.1 Environmental Peacebuilding: Emergence, Consolidation, and Evaluation

In the past two decades, the peace–environment nexus has crystalized as a distinct field of scholarship and practice, rooted in recognition that ecological and resource dynamics shape both conflict and security (Ali, 2007; Conca & Dabelko, 2002; Dresse et al., 2019; Swain & Öjendal, 2018). Defined as the sustainable management of natural resources to promote peace, the field highlights the environment's dual role: both a source of fragility and a platform for joint action (Matthew et al., 2009). The field connects to children, ethno-political conflict, forced migration (including disaster displacement), gender and social inequality, governance and institutional fragility, human rights claims, Indigenous peoples, livelihood insecurity and poverty, marginalized communities, public health impacts, and transboundary resource disputes, showing how ecological stress magnifies vulnerabilities.

The 1992 Rio Earth Summit laid crucial groundwork for environmental peacebuilding by institutionalizing sustainable development, natural resource governance, and global environmental cooperation. In line with this trajectory, the United Nations Environment Programme (UNEP) established a Post-Conflict and Disaster Management Branch in 1999—now incorporated into UNEP's Disasters and Conflicts Programme—to assess the environmental impacts of conflicts and disasters and confront the ensuing threats to human health, livelihoods, and security.

O. Serrat, *Advancing Environmental Peacebuilding*, SpringerBriefs in Political Science, https://doi.org/10.1007/978-981-92-4028-9_6

Around the same time, scholarship on environmental scarcity and violent conflict—most notably Homer-Dixon (1999)—provided the intellectual backdrop for these institutional developments. In 2001, the United Nations (UN) General Assembly declared November 6 as the International Day for Preventing the Exploitation of the Environment in War and Armed Conflict, signaling the normative recognition that environmental protection is inseparable from peace and security. This annual observance complements institutional developments by embedding environmental peacebuilding in global awareness and preventive practice. In 2005, the Environmental Peacebuilding Initiative was launched as a cross-regional collaboration among the Environmental Law Institute, McGill University, UNEP, and the University of Tokyo, further consolidating the field's policy and scholarly foundations.

Critics may argue that environmental factors are secondary to economic, political, or social drivers of conflict. However, in 2009, UNEP found that natural resources were implicated in 40% of intrastate conflicts since 1945,[1] yet also offered opportunities for dialogue, institution-building, and trust (UNEP, 2009). In 2010, in his report on peacebuilding in the aftermath of conflict, UN Secretary-General Ban Ki-moon urged Member States and UN agencies to acknowledge the importance of national ownership and access to natural resources in peacebuilding strategies (United Nations Secretary-General, 2010). Augmenting these efforts, the Environmental Law Institute, McGill University, UNEP, and the University of Tokyo jointly launched the Environmental Peacebuilding Platform in 2014. The Environmental Peacebuilding Platform expanded access to case studies, practitioner resources, and publications, and helped formalize the field's identity. These initiatives culminated in the incorporation of the Environmental Peacebuilding Association (EnPAx) in 2018, which gave the field consolidated institutional standing. Since then, EnPAx has expanded the Environmental Peacebuilding Platform into a broader global knowledge hub and community of practice (Environmental Peacebuilding Association, 2025). The field now enjoys both institutional acceptance and scholarly traction. This provides a foundation for methodological innovation.

Case studies exemplify that environmental peacebuilding rests on three guiding principles. Shared resource cooperation builds trust through the joint management of contested or transboundary resources. Sustainability in peace processes embeds ecological resilience into agreements and reconstruction efforts, ensuring that peace extends beyond immediate settlements. Inclusive natural resource governance prioritizes participation and equity, recognizing that marginalized groups must have a voice in resource decisions. Together, these principles embody the field's transformative potential. Yet practice unfolds within hybrid contexts—conditioned by

[1] This includes both high-value resources—such as minerals, oil, and timber—and essential livelihood assets like fertile land and water. Unlike intrastate conflicts, no widely accepted estimate exists for the share of interstate conflicts driven by natural resources. The difficulty lies in disentangling natural resources from overlapping drivers of interstate disputes (e.g., economic competition, geopolitical interests, governance fragility, historical grievances, identity politics), many of which are historically embedded and mutually reinforcing.

cultural, ecological, economic, governance, intergenerational, psychological, security, socio-political, and technological structures—that I frame in Appendix A to guide intervention design, implementation, and evaluation.

Google Trends (https://trends.google.com/; retrieved November 23, 2025), a website by Google that analyzes the popularity of top search queries in Google Search, reveals that worldwide interest in *environmental peacebuilding* as a topic remained negligible (0 points) from January 2004, the first date for which that website makes results available, until late 2016, when values began to rise into the 30–50 range, peaking at 49 in October 2017.[2] Interest continued through 2018–2019, reaching 55 in October 2019, and fluctuated between 17–39 points in 2022, with peaks in January and March. More recently, values stabilized in the 18–51 range during 2023–2024, before climbing sharply in 2025 to a peak of 69 in September and a maximum of 100 in November. These increases coincide with milestones such as UNEP's (2009) report on natural resources and conflict, the incorporation of EnPAx in 2018, and heightened global debate on climate–security linkages in the early 2020s. While the topic has not yet reached mainstream visibility comparable to other global concepts, its upward trajectory indicates growing acceptance of the role of climate adaptation, ecological cooperation, and natural resource governance in preventing conflict and sustaining peace.

Peace without proof is fragile. Unless environmental peacebuilding efforts are systematically evaluated, they risk becoming hopeful experiments rather than transformative pathways to resilience and reconciliation. Despite growing recognition and institutional consolidation, evaluative frameworks remain underdeveloped and offer limited guidance for assessing interventions. Jean and Ernstorfer (2014) highlight the lack of methodological consensus in complex, conflict-affected contexts, urging evaluations to capture contributions to "peace writ large" rather than confining analysis to project-level outputs. Paffenholz (2016) points to the implicit nature of theories of change and the absence of clear baselines, which weaken the rationale for how interventions are expected to generate impact. The Environmental Law Institute's (2023) *Toolkit on Monitoring and Evaluation of Environmental Peacebuilding* acknowledges the Organisation for Economic Co-operation and Development's Development Assistance Committee (OECD-DAC) six evaluation criteria—relevance, coherence, effectiveness, efficiency, impact, sustainability—but stresses that this framework is too narrow to encompass the dual objectives of environmental recovery and peacebuilding. It overlooks conflict dynamics, ecological resilience, and long-term social cohesion. Consequently, the Toolkit recommends adapting and supplementing the OECD–DAC framework with context-specific measures emphasizing adaptive learning, conflict sensitivity, equity, and inclusivity. Yet despite these advances, its guidance ultimately stops short of offering detailed methodological prescriptions, leaving practitioners with broad guidelines rather than concrete evaluative tools.

[2] Google Trends values are relative indices scaled so the highest point in the selected period equals 100; according to Google, a topic is a group of terms that share the same concept in any language, while a search term only includes data for that language.

To help fill this gap, this précis isolates three principles of environmental peacebuilding as a baseline for evaluation. Building on this foundation, its central contribution is to broaden the OECD-DAC evaluation framework with supplementary, peacebuilding-specific criteria, illustrated through comparative case analysis—a method well suited to capturing context-sensitive variation across ecological and political settings (OECD–DAC, 2019; Yin, 2018). Because the OECD–DAC's evaluation criteria are globally applied by UN agencies, donors, financial institutions, nongovernmental organizations, and national evaluation systems, they remain the most influential framework in international development. Therefore, this précis also situates them within hybrid environmental peacebuilding demands, identifying both baseline relevance and necessary adaptations. By positioning these measures as complementary rather than competing, this précis provides stakeholders with tools that enhance effectiveness, legitimacy, and responsiveness, even in the absence of OECD–DAC endorsement. Finally, this précis promotes a proactive orientation, framing environmental peacebuilding as a continuum in which anticipatory interventions prevent escalation and cultivate shared futures, rather than merely reacting to conflict. In sum, interventions prove most effective when anchored in clear principles, assessed with tailored criteria, and pursued through proactive pathways.

6.2 Principles: Operational Foundations of Environmental Peacebuilding

Where securitization narrows responses to ecological stress, environmental peacebuilding advances governance that not only alleviates immediate tensions but also reshapes conflict dynamics over time. This reorientation expands the scope of peacebuilding. Drawing on case studies, this précis distills three principles—shared resource cooperation, sustainability within peace processes, and inclusive natural resource governance—that anchor practice. Their effectiveness, however, is contingent on context: initial conditions, institutional arrangements, and desired outcomes determine when, how, and why conflict parties pursue cooperation. Even so, these principles provide operational baselines that are revisited and extended here. The case studies underscore variation across ecological systems, political settings, and governance structures, illustrating how environmental peacebuilding adapts under diverse conditions.

6.2.1 Operationalizing Shared Resource Cooperation

Shared ecological challenges can open platforms for cooperation across conflict lines. When states or communities depend on the same resource system, they must negotiate rules to manage competition. The Indus Waters Treaty (1960) between

India and Pakistan has endured through wars, showing how institutionalized cooperation can outlast geopolitical volatility and provide durable mechanisms for dispute resolution and technical collaboration (Dresse et al., 2019). In the Jordan River Basin, despite recent escalations in the Israel–Palestinian conflict (e.g., the Israel–Hamas war), EcoPeace Middle East—founded in 1994 and known for its Good Water Neighbors project (2001)—nurtured dialogue among Israeli, Jordanian, and Palestinian stakeholders through joint water management, showing how environmental interdependence can sustain cooperation even amid active conflict (Conca & Dabelko, 2002). Taken together, these cases show that cooperative governance mitigates scarcity while cultivating confidence and institutional habits of collaboration—even in adversarial contexts. Yet treaties may entrench inequities rather than resolve them, and cooperation can sometimes mask deeper disputes instead of transforming them.

6.2.2 *Institutionalizing Sustainability in Peace Processes*

Extending from shared resource cooperation, sustainability provisions decide whether peacebuilding gains last. Embedding ecological sustainability in peace agreements and reconstruction is essential to preventing relapse into violence. In Liberia, revenues from "conflict timber" fueled civil wars (1989–1996; 1999–2003) until timber sanctions by the United Nations (2003) and the U.S.-led Liberia Forest Initiative (2004) rebuilt governance and encouraged sustainable forestry, showing how sustainability measures can transform resource economies from conflict drivers into peacebuilding assets (Matthew et al., 2009; UNEP, 2009). In Colombia, the Peace Accord (2016) addressed land restitution, rural development, and environmental protection, incorporating sustainability within the peace architecture and acknowledging land grievances as structural drivers of conflict (Swain & Öjendal, 2018). These examples show that sustainability is not an add-on but a foundational prerequisite for durable peace.

6.2.3 *Embedding Inclusive Natural Resource Governance*

Unequal resource access and exclusion from decision-making drive conflict, while inclusive natural resource governance builds legitimacy and resilience. In Nepal, community forestry—initiated in the late 1970s and revitalized after the 2006 Comprehensive Peace Accord—empowered local groups to manage forests, delivering ecological gains and building social capital that strengthened post-conflict recovery and legitimacy (Ali, 2007). In Kenya, participatory resource management expanded after the 2007 presidential election and the ensuing 2007–2008 post-election violence, illustrating how participatory governance frameworks can defuse land and water disputes and foster trust in fragile political environments. By strengthening ecological outcomes and redressing grievances, inclusive governance reduces instability (Swain & Öjendal, 2018). However, research points to risks of elite capture and uneven benefit distribution, cautioning that inclusivity is not guaranteed by design alone (Platteau, 2004).

These three principles collectively ground environmental peacebuilding. Cooperation is most effective when embedded in institutions, sustainability provisions add greatest value when integrated early, and inclusivity consistently bolsters legitimacy and resilience. Yet principles alone cannot guarantee effectiveness; evaluation determines whether they translate into durable outcomes. To assess this, evaluative criteria must reflect hybrid contexts—ecological, political, and social forces shaping intervention design, implementation, and outcomes.

6.3 Criteria: Evaluating Practice and Outcomes

Peacebuilding thrives when learning is tested by evaluation—turning principles into practices that shape lasting outcomes for people and environments (Serrat, 2010, 2011, 2017a). Having identified three principles of environmental peacebuilding, the next challenge is to assess their effectiveness in real-world settings. Evaluative thinking entails ongoing reflection and adaptation, using critical inquiry to interpret evidence (Environmental Law Institute, 2023). It surfaces assumptions, probes questions, deepens understanding through reflection and perspective-taking, and informs decisions that lead to action (Serrat, 2017b).

Evaluating environmental peacebuilding requires frameworks that gauge ecological and peacebuilding outcomes with equal rigor. This mandate—linking environmental resilience and conflict transformation—demands evaluative tools that go beyond conventional development metrics. The evaluation criteria drawn by OECD–DAC—relevance, coherence, effectiveness, efficiency, impact, sustainability—remain the global standard, shaping judgments across international development (Organisation for Economic Co-operation and Development, 2019). Yet their uptake in environmental peacebuilding has been uneven and contested, in part owing to the field's dual ecological–peacebuilding mandate, which stretches the framework beyond its original design. As a result, evaluations, where they have been conducted, tend to privilege short-term relevance and coherence while neglecting efficiency, sustainability, and long-term impact in fragile, resource-dependent contexts (Serrat, 2025a, 2025b).

Extending these limitations, scholars have reinforced the concern. Zelli and Krause (2025) assert that evaluations must also account for how environmental recovery and livelihood security influence peacebuilding trajectories. Without careful contextualization, therefore, the OECD–DAC framework risks imposing development-centric logics on environmental peacebuilding. This précis extends the critiques by advancing supplementary criteria, tailored to hybrid contexts and informed by systems thinking (Serrat, 2025c).

6.3.1 *Six Drivers of Context*

Six contextual drivers—often outside the direct influence of environmental peacebuilders, yet still open to partial engagement—influence how the OECD–DAC framework is applied:

- **Community Engagement.** Genuine participation—not token consultation—builds legitimacy. Resolving power asymmetries (e.g., class, ethnicity, gender) fosters durability, ownership, and trust. This is especially vital for displaced and refugee communities, whose exclusion from decision-making perpetuates cycles of marginalization.
- **Conflict Intensity.** Conflict intensity dictates whether urgent stabilization or gradual trust-building is possible. Interventions must adapt as conditions shift.
- **Cultural Factors.** Cultural factors matter: respect for local customs and knowledge facilitates uptake, but culture is plural and contested. Balancing traditions with human rights and sustainability prevents romanticization and broadens legitimacy.
- **Economic Conditions.** Economic hardship fuels grievances, while inclusive opportunities generate peace dividends. Macro-level stability and local livelihoods shape feasibility.
- **Environmental Degradation.** Environmental degradation intensifies urgency and scope. Because degradation and conflict reinforce one another, ecological recovery must serve as both a peacebuilding goal and a means.
- **Political Stability.** Political stability enables implementation, while instability constrains it. Even in fragile, resource-dependent contexts, hybrid arrangements with civil society or traditional actors can provide entry points.

In practice, the six contextual drivers condition how the OECD–DAC framework operates in environmental peacebuilding: for instance, political instability undermines sustainability, while community engagement enhances relevance and impact.

6.3.2 *Adaptation: Extending the OECD–DAC Framework*

The six contextual drivers show that environmental peacebuilding unfolds not as a controlled intervention but as a negotiation with shifting ecological, political, and social forces. Therefore, the evaluation of environmental peacebuilding cannot be reduced to standard development metrics. Table 6.1 clarifies how contextual drivers recalibrate the relative weight of OECD–DAC's evaluation criteria.

As Table 6.1 shows, community engagement, cultural factors, environmental degradation, and political stability consistently rank high across all six evaluation criteria, underscoring their centrality to both immediate relevance and long-term sustainability. By contrast, conflict intensity and economic conditions heighten the salience of effectiveness and impact, while constraining coherence, efficiency,

Table 6.1 Influence of Contextual Drivers on the Relative Weighting of OECD–DAC's Evaluation Criteria in Environmental Peacebuilding

	Community Engagement	Conflict Intensity	Cultural Factors	Economic Conditions	Environmental Degradation	Political Stability
Relevance	High	Medium	High	Medium	High	High
Coherence	High	Medium	High	Medium	High	High
Effectiveness	High	High	High	Medium	High	Medium
Efficiency	High	Medium	Medium	High	Medium	Medium
Impact	High	High	High	High	High	High
Sustainability	High	Medium	High	Medium	High	High

relevance, and sustainability. This variation demonstrates the need for context-sensitive application, requiring evaluators to adapt criteria in fragile settings and to prioritize dimensions most critical to durable peace. In Appendix B, consistent with OECD–DAC's guidance on applying evaluation criteria thoughtfully (Organisation for Economic Co-operation and Development, 2021), I chart the baseline applicability of OECD-DAC's evaluation criteria to environmental peacebuilding and specify the adaptations required for hybrid contexts. Such refinements enable evaluations to move beyond technical measures, situating environmental peacebuilding within the broader global agenda[3] while reinforcing relevance to long-term sustainability and durable peace.

6.3.3 *Methodological and Structural Challenges*

Environmental peacebuilding requires evaluative criteria that move beyond the OECD-DAC framework, which—though valuable in international development—remains insufficiently tailored to the field's hybrid nature. Without criteria attuned to conflict dynamics, ecological systems, and governance legitimacy, evaluations risk both technical blind spots and normative distortions.

Methodological hurdles expose this absence most clearly. Attribution is difficult because interventions intertwine with political and social dynamics (Bush, 1998; Environmental Law Institute, 2023). As Carius (2002) observes, methods for assessing transboundary environmental projects remain insufficiently tested, while outcomes often unfold over extended timeframes and across wide geographical areas, complicating short-term assessment (Environmental Law Institute, 2023). These systemic blind spots demand longitudinal methods with adaptive indicators—criteria capable of tracking ecological resilience, social change, and legitimacy over time. The challenge intensifies when interventions intersect with ethno-political

[3] Within the global agenda, environmental peacebuilding is most notably reflected in the Sustainable Development Goals (SDGs), particularly SDG 16 (Peace, Justice, and Strong Institutions) and SDG 13 (Climate Action) as presented in *Transforming Our World: The 2030 Agenda for Sustainable Development* (United Nations, 2015).

cleavages or forced migration, where causality becomes elusive. Attribution remains complex, but the aim should be to capture contributions and trajectories rather than isolate single causes. In such contexts, evaluators increasingly rely on mixed-methods approaches that combine conflict event data with stakeholder perspectives, bolstering validity and revealing layered complexity—especially when informed by disciplines such as anthropology, economics, environmental science, history, international relations, and sociology (Serrat, 2025d, 2025e). Emerging tools such as geospatial analysis and remote sensing further enhance ecological monitoring, but their integration into peacebuilding evaluation remains experimental and not yet systematically established. Their value depends on criteria that explicitly link environmental change to conflict dynamics.

Structural critiques likewise underscore the need for tailored criteria. Technical effectiveness alone cannot guard against normative risks such as masking power asymmetries or sidelining environmental concerns in fragile states. Cooperative initiatives may deflect attention from disputes rather than resolve them, while entrenched inequities in resource access persist (Ali, 2007; Conca & Dabelko, 2002). Similarly, fragile states often prioritize immediate security imperatives over environmental concerns (Ali, 2007; Conca & Dabelko, 2002). Criteria that foreground equity, legitimacy, and ecological sustainability are therefore essential to ensure cooperation is not merely instrumental but transformative.

Together, these methodological and structural challenges reveal that the absence of tailored evaluative criteria is not a peripheral issue but the central obstacle to rigorous and meaningful assessment. Advancing the field requires criteria that integrate technical rigor with normative sensitivity—anchoring evaluation in conflict-sensitive, ecologically resilient, and equity-driven measures capable of bridging scholarship and practice.

6.3.4 Supplementary Evaluative Criteria

The OECD–DAC framework was not intended to account for the interplay of ecological dynamics, conflict transformation, and governance legitimacy that define environmental peacebuilding—a gap long noted in peace and conflict impact assessments (Bush, 1998). Supplementary evaluative criteria are needed to transcend technical measures, reflecting the hybrid contexts of environmental peacebuilding. I propose seven supplementary evaluative criteria,[4] each targeting a distinct gap and enhancing relevance to complex, cross-sectoral settings (Serrat, 2025f):

[4] The supplementary evaluative criteria I propose draw on firsthand evaluation experience in Asia and the Pacific and on sustained critiques of the OECD-DAC framework when applied to the hybrid contexts of environmental peacebuilding. Their relevance must be tested through in-depth case studies.

- **Adaptability and Resilience.** This criterion articulates the capacity of interventions to adjust to shifting ecological and social conditions, ensuring that peacebuilding efforts remain relevant and effective even as contexts evolve.
- **Conflict Sensitivity and "Do No Harm".** Evaluations should consider whether environmental peacebuilding interventions avoid heightening tensions and instead contribute to reducing the risk of renewed conflict.
- **Cultural Sensitivity.** Effective peacebuilding requires respect for and integration of local knowledge, traditions, and practices, while balancing these with broader principles of sustainability and human rights.
- **Inclusivity and Participation.** Interventions gain strength when they engage marginalized groups— displaced populations, Indigenous peoples, minoritized communities, women, youth—while ensuring fair benefit-sharing to build legitimacy and ownership. Extending inclusivity also means recognizing children's vulnerabilities, from education to health.
- **Legitimacy.** Building trust depends on accountability, transparency, and alignment with local norms and expectations, which together foster durable support for environmental governance.
- **Long-Term Environmental Impact.** Beyond immediate outcomes, evaluations should assess the enduring ecological benefits of interventions, recognizing that sustainable peace is inseparable from environmental resilience.
- **"Peace Writ Large".** Interventions should be judged by their contributions to broader peace and stability, including their influence on national or regional trajectories. This signals systemic impacts that transcend immediate project outcomes.

These supplementary evaluative criteria add conflict-sensitivity, ecological depth, local dimensions (e.g., community perceptions, cultural practices, livelihood dependencies), and legitimacy that the OECD–DAC framework does not fully reflect (Ide, 2024; Serrat, 2025a, 2025b, 2025f).[5] They are not optional add-ons but indispensable lenses without which evaluation cannot grasp the field's hybrid nature. They navigate the methodological and structural challenges outlined earlier, offering a more holistic lens for assessing environmental peacebuilding interventions. By embedding peacebuilding values and transformative ambitions, they ensure assessments encompass ecological resilience and legitimacy. Table 6.2 shows how these criteria interact with contextual drivers, underscoring their peacebuilding relevance.

Table 6.2 complements the OECD–DAC framework with peacebuilding-specific evaluative lenses: adaptability and resilience, conflict sensitivity and "do no harm", cultural sensitivity, inclusivity and participation, legitimacy, long-term environmental impact, and "peace writ large". Consistent "High" ratings across the community

[5] Depending on context, scholars and practitioners may contend that additional evaluative criteria—such as justice and equity, scalability and replicability, institutional capacity, and climate responsiveness—are warranted. In the interest of parsimony, however, the supplementary evaluative criteria proposed here are framed to subsume these concerns, while acknowledging they may merit explicit treatment in certain contexts.

Table 6.2 Applicability of Supplementary Evaluative Criteria Across Environmental Peacebuilding Contextual Drivers

	Community Engagement	Conflict Intensity	Cultural Factors	Economic Conditions	Environmental Degradation	Political Stability
Adaptability & Resilience	High	High	High	Medium	High	High
Conflict Sensitivity & "Do No Harm"	High	High	High	Medium	High	High
Cultural Sensitivity	High	Medium	High	Medium	High	High
Inclusivity & Participation	High	Medium	High	Medium	High	High
Legitimacy	High	Medium	High	Medium	High	High
Long-Term Environmental Impact	High	Medium	High	Medium	High	High
"Peace Writ Large"	High	High	High	Medium	High	High

engagement, cultural factors, environmental degradation, and political stability dimensions highlight their role as core pillars of environmental peacebuilding. "Medium" ratings in the conflict intensity dimension suggest that cultural sensitivity and inclusivity are constrained by polarized narratives, mistrust, or limited civic space, while legitimacy and long-term impact suffer from short-termism and institutional fragility. Recurring "Medium" ratings under the economic conditions dimension suggest that although economic fragility may limit application, it does not diminish the importance of the supplementary evaluative criteria.

Tables 6.1 and 6.2 demonstrate that while the OECD–DAC framework offers a foundation—further refined in Appendix B—supplementary evaluative criteria are essential to encompass the full scope of environmental peacebuilding's transformative potential. With this evaluative architecture in place, the next step is to explore how principles and criteria can be applied proactively—before conflict escalates.

6.4 Designing Ahead: Operationalizing Proactive Environmental Peacebuilding

Through evaluation, both outcomes and pathways for proactive strategy come into view. While environmental peacebuilding interventions often follow violence and remain constrained by mistrust and weak institutions, anticipatory approaches identify ecological drivers early and integrate cooperative mechanisms before disputes intensify. The 2012 UNEP report *Greening the Blue Helmets*—the first to link natural resources directly to peacekeeping—affirmed the importance of integrating environmental considerations into preventive diplomacy (UNEP, 2012).

6.4.1 Proactive Shared Resource Cooperation

In the Great Lakes Region of Africa, UNEP and the International Conference on the Great Lakes Region established in 2004 have sought to curb cross-border resource exploitation and promote cooperative management of forests and minerals (Matthew et al., 2009). These efforts illustrate how resource interdependence can be harnessed for preventive diplomacy before disputes reignite, though persistent displacement and insecurity expose the limits of cooperation without broader institutional support. In the Great Lakes region, however, cross-border displacement and refugee movements signal how resource governance, migration, and security are tightly interwoven, making proactive cooperation doubly urgent.

6.4.2 Proactive Sustainability in Peace Processes

In Colombia, environmental provisions were incorporated in the 2016 Peace Accord itself (Swain & Öjendal, 2018). These provisions not only enshrined land restitution but also recognized the ethno-territorial rights of Indigenous and Afro-Colombian communities, linking sustainability to justice and human rights. By including commitments to rural reform, land restitution, and sustainable development at the outset, Colombia demonstrated how proactive sustainability design can align environmental justice with long-term peacebuilding goals.

6.4.3 Proactive Inclusive Natural Resource Governance

The Arctic Council constituted in 1996 provides a compelling model: its founding structure granted six Indigenous Peoples' organizations "Permanent Participant" status, institutionalizing inclusivity as a preventive peacebuilding tool and embedding marginalized voices in environmental governance (Ali, 2007). This institutional design has helped prevent disputes over Arctic resources from escalating, demonstrating inclusivity as a preventive peacebuilding tool, though sovereignty disputes and resource competition continue to test the Council's preventive capacity.

Taken together, these case studies elucidate how environmental peacebuilding principles can be operationalized to anticipate rather than merely respond to conflict. However, proactive interventions are not without risks: contested resource claims, institutional fragility, and political resistance may constrain uptake. These challenges must be identified and confronted in future designs. Proactive interventions cultivate resilience by embedding cooperation, sustainability, and inclusivity before ecological stressors escalate. By shifting orientation from reactive to proactive, environmental peacebuilding moves closer to fulfilling its transformative potential. Yet concerns remain that without systemic economic reforms, accelerating climate change will overwhelm environmental peacebuilding by intensifying ecological stressors.

6.5 Future Directions

Environmental peacebuilding must transcend *ad hoc* responses to operationalize shared resource cooperation, institutionalize sustainability in peace processes, and embed inclusive natural resource governance into the architecture of environmental peacebuilding. Doing so is vital if, as the *White Paper on the Future of Environmental Peacebuilding* urges, we are to cultivate an infrastructure for peace—a coherent system of agreements, institutions, and practices that foster cooperation and strengthen resilience across ecological and social domains (Geneva Peacebuilding Platform et al., 2022).

The *White Paper on the Future of Environmental Peacebuilding* (Geneva Peacebuilding Platform et al., 2022) also identifies building and sharing a robust evidence base as a central agenda item. It affirms that "monitoring, evaluation, and knowledge-sharing are indispensable for demonstrating impact, ensuring accountability, and integrating environmental peacebuilding within education, funding, and governance systems" (p. 14). This call directly underpins the need to refine evaluation frameworks and formulate supplementary evaluative criteria.

Three priorities stand out, each reinforcing the need for evaluative rigor and institutional uptake:

- **Refining Evaluation Frameworks Through Comparative Case Analysis and Longitudinal Methods.** Future research should expand comparative case analysis, track long-term impacts through longitudinal studies, and refine evaluative tools that align ecological and peacebuilding outcomes. This means both adapting the OECD–DAC framework and supplementing it with peacebuilding-specific evaluative criteria (Serrat, 2025a, 2025b, 2025f).
- **Advancing Supplementary Criteria Tailored to Hybrid Ecological and Political Contexts.** Research should ask a central question: How can evaluative criteria be tailored to the ecological, resource-management, and socio-political challenges that shape diverse environmental peacebuilding contexts? Subquestions include: (1) What are the limitations of the OECD–DAC framework, and how might it be adapted to better address the hybrid contexts of environmental peacebuilding? (2) What supplementary evaluative criteria best address the layered challenges of environmental peacebuilding? (3) To what extent do the conditions and dynamics of existing case studies suggest the relevance and contextual fit of newly formulated evaluative criteria?
- **Bridging Scholarship and Practice by Embedding Insights into Institutional Commitments.** A robust evidence base is essential—not only for demonstrating impact and ensuring accountability, but also for embedding environmental peacebuilding into education, funding, and governance systems. For policymakers and practitioners, the challenge is to translate insights into institutional commitments—embedding environmental peacebuilding within funding frameworks, peace agreements, and governance mechanisms for conflict prevention (Geneva Peacebuilding Platform et al., 2022).

Comparative case analysis offers a rigorous method for deriving evaluative criteria that are both generalizable and adjustable, showcasing recurring patterns as well as context-specific approaches that can inform future design (Serrat, 2025f; Yin, 2018).[6] George and Bennett (2005) highlight how case studies contribute to theory development. Indicative case studies—from the Amazon in South America, the Heart of Borneo in Southeast Asia (Serrat, 2025g), the Indus Waters Treaty in South Asia, Lake Chad in Africa (Serrat, 2025h), the Mekong River spanning multiple Asian states (Serrat, 2004, 2025i), and climate resilience initiatives in the Caribbean with conflict-sensitive dimensions—demonstrate not only the breadth of environmental peacebuilding challenges across lakes and rivers, forests and watersheds, coastal zones and small island states, but also their geographical spread across diverse world regions. Together, they reveal the uneven landscape of cooperation, ranging from enduring frameworks to fragile arrangements. To support this forward-looking agenda, I outline in Appendix C a methodological framework for structuring case study evaluations, providing practical guidance tailored to environmental peacebuilding and a sample semi-structured interview guide. In Appendix D, I present a flow diagram of six key stages of an environmental peacebuilding case study and a layered template organized into seven sections for documenting interventions.

The stakes are escalating: climate change is intensifying resource pressures, driving environmental migration, and straining governance. Environmental peacebuilding has matured into a recognized field, yet evaluation remains its most deficient dimension. Methodological hurdles—such as attribution problems, temporal complexity, and the difficulty of tracing ecological–conflict linkages—combine with structural constraints like donor asymmetries, institutional fragmentation, and the sidelining of environmental concerns in fragile states. Without more rigorous evaluation, interventions risk rhetorical celebration without evident impact.

The supplementary criteria advanced here—adaptability and resilience, conflict sensitivity and "do no harm," cultural sensitivity, inclusivity and participation, legitimacy, long-term environmental impact, and "peace writ large"—offer a pathway beyond development-centric logics toward holistic, context-sensitive assessment. By embedding the supplementary criteria into evaluation, the field can aspire to transform its weakest link into a stronger asset. Interdisciplinary, participatory, and proactive evaluation is not simply a technical exercise but a transformative practice: it enables institutions to anticipate ecological stressors, align sustainability with justice, and institutionalize environmental peacebuilding into donor frameworks and governance systems. In an era of accelerating climate pressures and fragile political orders, the decisive test is whether evaluation can evolve swiftly enough to convert environmental peacebuilding from aspiration into durable, measurable peace. This précis advocates precisely such research, advancing principles, adaptations to the OECD–DAC framework, and supplementary evaluative criteria as the foundation for that evolution.

[6] Yin (2018) notes that case study research is ideally appropriate for examining complex, real-world interventions where the boundaries between phenomenon and context are blurred, making it well-suited to environmental peacebuilding.

Appendix A

See Fig. 6.1

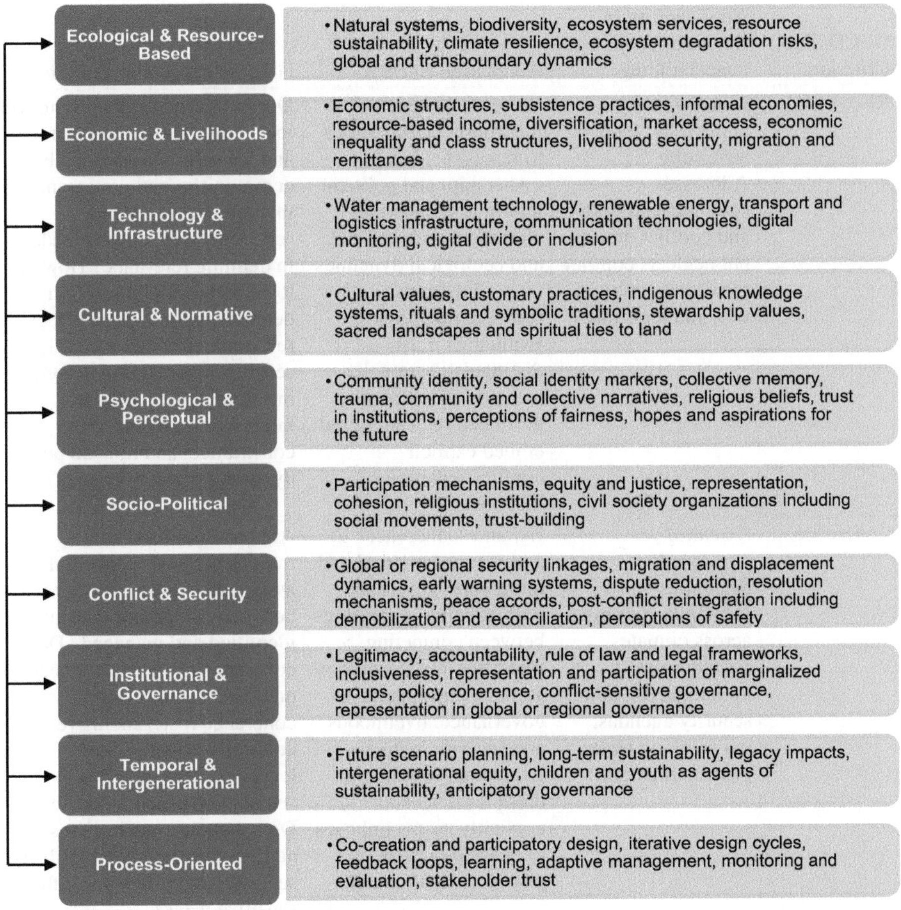

Fig. 6.1 Hybrid contextual framework for designing, implementing, and evaluating Environmental Peacebuilding interventions. Note. This framework sequences ecological systems, economic livelihoods, technological infrastructures, cultural and normative values, psychological and perceptual dimensions, socio-political relations, conflict and security dynamics, and institutional governance structures, while also integrating temporal and intergenerational sustainability and process-oriented learning. Together, these domains highlight the layered intersections that should inform the design, implementation, and evaluation of environmental peacebuilding interventions

Appendix B

See Table 6.3

Table 6.3 Adaptation requirements for OECD–DAC's evaluation criteria in environmental peacebuilding contexts

OECD–DAC Criterion	Applicability in Environmental Peacebuilding	Adaptation Needed	Guiding Evaluative Questions
Relevance	*Conditionally applicable, requires reframing.* Addresses ecological stressors and conflict drivers but neglects conflict sensitivity, cultural ties, and legitimacy.	Reframe "relevance" to incorporate cultural ties to resources, local ownership, and political legitimacy, ensuring alignment with conflict and ecological dynamics (e.g., climate risks, resource pressures). Evaluators must distinguish externally imposed vs. locally grounded relevance, and embed explicit conflict-sensitivity diagnostics.	Does relevance explicitly reflect local cultural ties to resources and legitimacy as defined by communities and governance institutions? Is conflict sensitivity explicitly embedded in defining relevance? How is local ownership ensured in determining relevance? Does relevance strengthen conflict-sensitive cooperation and equity in resource use? Whose interests are advanced or constrained, and how do power dynamics affect relevance?
Coherence	*Essential yet structurally demanding.* Demands alignment across climate, development, humanitarian, and security agendas, but weak on peace process linkages.	Expand "coherence" to assess cross-sectoral and multi-level governance, with attention to tensions between competing mandates and integration across environment, governance, livelihoods, and security sectors. Evaluators should map these tensions and explicitly assess linkages to peace processes, not just sectoral alignment.	Are cross-sectoral and multi-level governance linkages assessed? How are tensions between competing mandates identified and managed? Does coherence explicitly connect to peace processes? Does coherence foster inclusive, culturally sensitive, and legitimate integration across sectors and peace processes? Do coherence mechanisms address transboundary scale and security actor roles in conflict-sensitive ways?

(continued)

Table 6.3 (continued)

OECD–DAC Criterion	Applicability in Environmental Peacebuilding	Adaptation Needed	Guiding Evaluative Questions
Effectiveness	*Partially adequate, risks reductionism.* Measures outcomes but risks overlooking relational gains such as trust, cooperation, and broader peace contributions.	Use dual metrics to track conflict transformation and ecological resilience, ensuring "effectiveness" is not reduced to technical outputs alone. Add explicit indicators for relational outcomes (e.g., cooperation, joint resource management, trust) alongside ecological gains.	Are ecological resilience and conflict transformation outcomes measured in ways that are culturally appropriate and perceived as legitimate by local actors? Do indicators integrate relational gains? Is effectiveness assessed beyond technical outputs? Does effectiveness capture adaptive and peacebuilding capacities in conflict contexts? Are do-no-harm baselines, counterfactuals, and safeguards used to prevent harm and substantiate causal claims?
Efficiency	*Problematic without redefinition.* Conventional cost–benefit logic undervalues inclusive, time-intensive processes and ignores adaptive management.	Redefine "efficiency" to recognize that slower, participatory approaches may be more efficient in sustaining peace long-term, while also considering the cost-effectiveness of joint environmental and peace outcomes. Evaluators should weigh long-term peace dividends against short-term costs, and include adaptive management as a criterion of efficiency.	Does efficiency account for participatory, culturally appropriate, and legitimate processes, even when time-intensive? Are long-term peace dividends weighed against short-term costs? Is adaptive management included as part of efficiency? Does efficiency balance resource use with equity and conflict-sensitive participation? Are adaptive learning triggers and decision adjustments documented to improve efficiency over time?
Impact	*Relevant but insufficiently integrated.* Tracks long-term effects but struggles to integrate ecological, political, and social dimensions, and neglects inclusivity and equity.	Disaggregate "impact" across environmental, governance, and social domains, and develop methods to integrate their interdependence. Add explicit attention to equity and inclusivity, ensuring beneficial impacts on marginalized groups are visible, and encourage use of resilience metrics or systems mapping to illuminate integration.	Is impact disaggregated across environmental, governance, and social domains? Do methods capture interdependence among these domains? Are equity, inclusivity, and cultural legitimacy explicitly assessed, especially for diverse cultural communities and marginalized groups? Does impact demonstrate conflict transformation and equitable benefit-sharing? Are unintended consequences, spillovers, and differential impacts systematically assessed?

(continued)

Table 6.3 (continued)

OECD–DAC Criterion	Applicability in Environmental Peacebuilding	Adaptation Needed	Guiding Evaluative Questions
Sustainability	*Critical but conceptually underdefined.* Aligns with ecological resilience and durable peace, but weak on values, principles, and institutional durability.	Requires significant adaptation of "sustainability," including integration of ecological thresholds, governance capacity, and intergenerational equity, while also assessing the durability of cooperation mechanisms, environmental agreements, and institutions. Evaluators should measure resilience across ecological tipping points, generations, and institutional values and principles underpinning durable peace.	Are ecological thresholds and tipping points integrated into assessments of sustainability? Is governance capacity and institutional durability evaluated? Does sustainability include intergenerational equity, cultural continuity, and legitimacy of institutions underpinning durable peace? Does sustainability embed conflict-sensitive cooperation and institutional resilience across generations? Are cooperation mechanisms sequenced to reduce legacy grievances and embed accountability across generations?

Note. The OECD-DAC's evaluation criteria frame inquiry around core questions: Is the intervention doing the right things (see Relevance)? How well does the intervention fit (see Coherence)? Is the intervention achieving its objectives (see Effectiveness)? How well are resources being used (see Efficiency)? What difference does the intervention make (see Impact)? Will the benefits last (see Sustainability)? This table situates the OECD-DAC's evaluation criteria within the demands of hybrid environmental peacebuilding, identifying both their baseline relevance and the adaptations required. Although expanded guiding questions may appear overwhelming, they function as flexible lenses that evaluators can tailor to context. By sharpening the focus of these six lenses, the table reinforces the central argument of this précis: meaningful evaluation must integrate ecological thresholds, governance capacity, and fairness in resource distribution alongside conventional technical measures. The proposed adaptations build on OECD–DAC's guidance on applying evaluation criteria thoughtfully (Organisation for Economic Co-operation and Development, 2021), which emphasizes tailoring the OECD–DAC's evaluation criteria to context and purpose, and OECD–DAC's guidance on evaluating peacebuilding activities in settings of conflict and fragility (Organisation for Economic Co-operation and Development, 2012), which recognized the limitations of standard evaluation in conflict-affected settings. Together, the five guiding questions under each criterion extend evaluation beyond technical outputs to capture the distinct demands of environmental peacebuilding, ensuring systematic attention to conflict sensitivity, inclusion, relational outcomes, adaptive learning, accountability, and unintended consequences. In doing so, the adapted criteria explicitly embed peacebuilding-specific evaluative lenses—resilience and adaptability, conflict sensitivity and do-no-harm safeguards, cultural sensitivity, inclusivity and participation, legitimacy, long-term environmental impact, and contributions to "peace writ large"

Appendix C

See Tables 6.4 and 6.5

Table 6.4 Framework for Structuring Case Study Evaluations in Environmental Peacebuilding

Research Design
• Frame the research question to foreground "how" or "why," and use subquestions that mix "what," "how," and "why." • Justify the connection between the research question (and subquestions) and the data to be gathered. • State the comparative logic of cross-case study comparisons. • Explain how this strategy links to the research question (and subquestions) and the overall research design.
Analytical Framework
• Describe the analytical approach (e.g., pattern-matching, explanation building). • Provide a rationale for the chosen analytical framework, including definitions and operationalizations of key concepts. • Define evaluative criteria. • Perform institutional analysis, integrating environmental governance and peacebuilding dimensions. • Incorporate triangulation and validity checks to strengthen credibility. • Reflect on researcher positionality and its potential influence on interpretation.
Case Study Selection
• Specify criteria for case study inclusion (e.g., conflict intensity, ecological diversity, governance structures). • Specify criteria for case study exclusion (e.g., data scarcity, lack of stakeholder access, ethical concerns). • Justify inclusion or exclusion decisions, noting whether recency enhances relevance or whether historical depth offers critical insights. • Define each case study by delineating its institutional, spatial, and temporal boundaries. • Map key stakeholders. • Situate selected case studies within the context of environmental peacebuilding relevance (e.g., ecological stressors, resource governance challenges). • Ensure comparability across case studies while preserving variation that illuminates causal mechanisms. • Make out the theory of change behind each case study, identifying expected causal pathways and assumptions underlying the intervention model.
Data Sources
• Review academic literature, newspaper articles, and official documents. • Compile and indicate sources of evidence (e.g., appended list of documents, profiles of available interviewees, where feasible). • Conduct semi-structured, open-ended, thematic interviews with stakeholders where feasible, while recognising that environmental peacebuilding interventions often span vast spatial and temporal continuums and that documentary sources may therefore provide the primary evidence base in many cases. • Assess the quality and reliability of sources, noting potential bias or framing effects. • Triangulate evidence across primary and secondary sources to strengthen validity. • Describe unexpected data collection barriers and explain their potential effects on findings.

(continued)

Table 6.4 (continued)

Limitations
• Acknowledge inherent shortcomings in the research design (e.g., limited number of case studies, narrow institutional focus). • Recognize analytical framework constraints (e.g., difficulty integrating multi-level dynamics, risk of reductionism). • Identify data limitations (e.g., restricted access to interviewees, reliance on documentary sources). • Discuss how these limitations may have shaped the findings. • Offer transparent caveats on methodological constraints specific to environmental peacebuilding. • Clarify the limited generalizability of results and the challenges of disseminating findings to practitioner audiences.

Note. This table proposes a methodological framework for evaluating case studies in environmental peacebuilding. It delineates five core components of a rigorous methodology—research design, analytical framework, case study selection, data sources, and limitations—to support transparent and consistent application of evaluative criteria. By anchoring case study methodology in ecological, governance, and social imperatives, the framework responds to critiques of under-tested and methodologically constrained evaluation practices in the field. To improve usability, it contributes guidance on framing research questions, prioritizing sources, triangulating evidence, and selecting analytical tools, thereby integrating conceptual depth with operational clarity

Table 6.5 Sample Interview Guide for Environmental Peacebuilding Interventions

Opening & Context
• Can you describe your role in relation to environmental peacebuilding in this context? • How have environmental issues (e.g., ecological stressors, resource scarcity) intersected with conflict or cooperation here?
Conflict–Environment Dynamics
• In your view, what are the main environmental stressors contributing to cooperation or tensions? • How do resource governance structures (formal or informal) shape conflict resolution and environmental peacebuilding efforts?
Institutional & Policy Dimensions
• What institutions (e.g., local, national, international) are most influential in managing environmental conflict here? • How effective have policies or agreements been in balancing ecological sustainability with environmental peacebuilding goals?
Stakeholder Perspectives
• Which actors (e.g., civil society, communities, government, nongovernment organizations, private sector) play the most significant roles in environmental peacebuilding? • How do power dynamics among stakeholders affect cooperation or exacerbate tensions?
Intervention Pathways
• What interventions (e.g., agreements, community initiatives, projects) have been attempted, and what outcomes have you observed? • What theory of change (or assumptions) underpinned these interventions, and were they realistic?
Resilience & Sustainability
• How have communities adapted to environmental stressors in ways that promote peace? • What factors contribute to resilience and sustainability in this context?
Reflections & Lessons
• What limitations or challenges have you encountered in environmental peacebuilding efforts? • What lessons would you highlight for future interventions in similar contexts?

Note. The semi-structured, open-ended, thematic questions in this table cover the full arc of environmental peacebuilding inquiry to ensure that research design, case selection, analytical choices, data use, and limitations are all addressed in an integrated way to produce credible, comparable, and context-sensitive findings

Appendix D

See Fig. 6.2 and Table 6.6

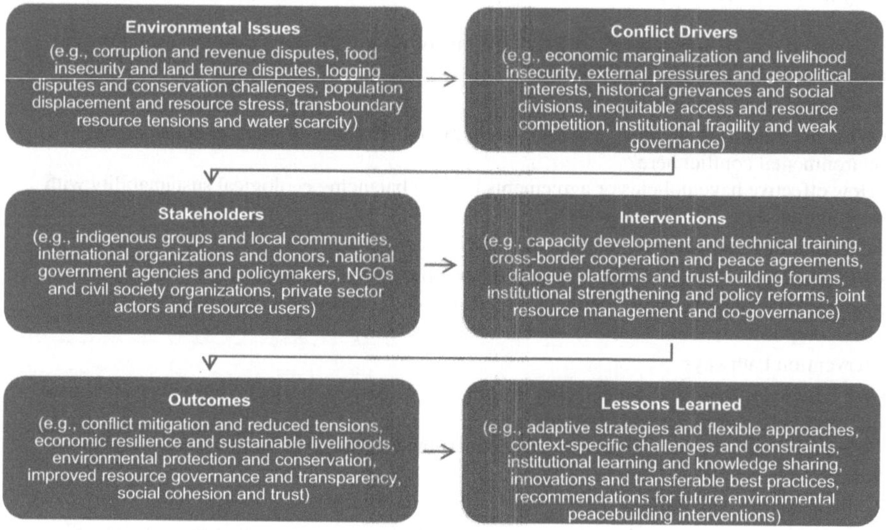

Fig. 6.2 Key Stages of an Environmental Peacebuilding Case Study. Note. This diagram illustrates six stages of an Environmental Peacebuilding case study: environmental issues, conflict drivers, stakeholders, interventions, outcomes, and lessons learned. The framework supports systematic analysis across local and transboundary cases. Arrows indicate causal progression, and shaded boxes highlight stakeholder engagement

Table 6.6 Layered Template for an Environmental Peacebuilding Case Study

Section	Core Components	Optional Enhancements
Introduction & Context	Environmental stressors; geographic setting; socio-political setting	Climate vulnerability profile; historical background; legal and policy context
Conflict Analysis	Actors; conflict dynamics; drivers of conflict	Conflict spoilers; conflict typology; peace enablers; structural vs. proximate causes; temporal evolution; trigger events
Stakeholder Engagement	Challenges to engagement; mapping; participation mechanisms	Gender and inclusion lens; power analysis; stakeholder incentives and disincentives; trust-building methods
Intervention & Strategy	Description of the intervention; implementation process; peacebuilding approach	Funding sources; monitoring and evaluation mechanisms; theory of change
Outcomes & Impacts	Environmental improvements; peacebuilding results; social benefits	Environmental indicators; equity and distributional outcomes; institutional changes; unintended consequences
Lessons Learned	Challenges; replicability; success factors	Adaptive learning; contextual limitations; missed opportunities; negative lessons; scalability conditions
Conclusion & Recommendations	Future directions; policy implications; summary insights	Comparative analysis; scenario modeling; tailored recommendations; visual aids

Note. This template offers a structured, adaptable format for documenting environmental peacebuilding interventions. It is designed to support both concise summaries and in-depth comparative analysis across diverse contexts. The template is organized into seven core sections, each with essential components and optional enhancements that allow for deeper exploration when needed

References

Ali, S. (Ed.). (2007). *Peace parks: Conservation and conflict resolution*. MIT Press.

Bush, K. (1998). *A measure of peace: Peace and conflict impact assessment (PCIA) of development projects in conflict zones*. International Development Research Centre.

Carius, A. (2002). *Environmental peacemaking*. Woodrow Wilson International Center for Scholars.

Conca, K., & Dabelko, G. (Eds.). (2002). *Environmental peacemaking*. Woodrow Wilson Center Press.

Dresse, A., Fischhendler, I., Nielsen, J., & Zikos, D. (2019). Environmental peacebuilding: Towards a theoretical framework. *Cooperation and Conflict, 54*(1), 99–119. https://doi.org/10.1177/0010836718808331

Environmental Law Institute. (2023). *Toolkit on monitoring and evaluation of environmental peacebuilding*. Environmental Law Institute. https://www.m-and-e.environmentalpeacebuilding.org/

Environmental Peacebuilding Association. (2025). *Environmental peacebuilding knowledge platform*. https://www.environmentalpeacebuilding.org/.

George, A., & Bennett, A. (2005). *Case studies and theory development in the social sciences*. MIT Press.

Geneva Peacebuilding Platform, Environmental Peacebuilding Association, Environmental Law Institute, & PeaceNexus Foundation. (2022). *White paper on the future of environmental peacebuilding*. Geneva Centre for Security Policy. https://www.gcsp.ch/publications/white-paper-future-environmental-peacebuilding

Homer-Dixon, T. (1999). *Environment, scarcity, and violence*. Princeton University Press.

Ide, T. (2024). From climate conflicts to environmental peacebuilding: Exploring local dimensions. *Environment and Security, 2*(1), 3–20.

Jean, I., & Ernstorfer, A. (2014). *Alternative approaches to evaluating peacebuilding initiatives*. ALNAP. https://alnap.org/help-library/resources/alternative-approaches-to-evaluating-peacebuilding-initiatives/

Matthew, R., Brown, O., & Jensen, D. (2009). *From conflict to peacebuilding: The role of natural resources and the environment*. United Nations Environment Programme.

Organisation for Economic Co-operation and Development. (2012). *Evaluating peacebuilding activities in settings of conflict and fragility: Improving learning for results*. OECD Publishing. https://doi.org/10.1787/9789264106802-en

Organisation for Economic Co-operation and Development. (2019). *Better criteria for better evaluation: Revised evaluation criteria definitions and principles for use*. OECD Publishing. https://doi.org/10.1787/543e84ed-en

Organisation for Economic Co-operation and Development. (2021). *Applying evaluation criteria thoughtfully*. OECD Publishing. https://doi.org/10.1787/543e84ed-en

Paffenholz, T. (2016). Methodologies in peacebuilding evaluation: Testing new approaches. In T. Paffenholz (Ed.), *Peacebuilding evaluation: Principles and practice* (pp. 1–24). Lynne Rienner Publishers.

Platteau, J. (2004). Monitoring elite capture in community-driven development. *Development and Change, 35*(2), 223–246.

Serrat, O. (2004). *A note on Mekong River management*. Unpublished manuscript, Asian Development Bank. https://www.researchgate.net/publication/291056409_A_Note_on_Mekong_River_Management

Serrat, O. (2010). *Learning in development*. Asian Development Bank. https://www.adb.org/publications/learning-development

Serrat, O. (2011). *Learning for peace* [Presentation]. Asian Development Bank. https://www.researchgate.net/publication/275346096_Learning_for_Peace

Serrat, O. (2017a). Learning from evaluation. In *Knowledge solutions: Tools, methods, and approaches to drive organizational performance* (pp. 879–884). Springer.

Serrat, O. (2017b). Critical thinking. In *Knowledge solutions: Tools, methods, and approaches to drive organizational performance* (pp. 889–896). Springer.

Serrat, O. (2025a). Harnessing evaluative criteria for environmental peacebuilding. In *Anthropogenic solutions for climate change: Achieving environmental peace* (pp. 171–208). Springer.

Serrat, O. (2025b). Environmental peacebuilding: Nurturing hope with analytical persuasion. In *Anthropogenic solutions for climate change: Achieving environmental peace* (pp. 165–170). Springer.

Serrat, O. (2025c). *Accounting for complexity: Liberal studies and environmental peacebuilding*. Unpublished manuscript, Georgetown University. https://www.researchgate.net/publication/396824279_Accounting_for_Complexity_Liberal_Studies_and_Environmental_Peacebuilding

Serrat, O. (2025d). Disciplinary perspectives on environmental peacebuilding: Insights from environmental science and international relations. In *Anthropogenic solutions for climate change: Achieving environmental peace* (pp. 147–164). Springer.

Serrat, O. (2025e). *Bridging the gap: Synergizing anthropology, economics, history, and sociology for environmental peacebuilding*. Unpublished manuscript, Georgetown University. https://www.researchgate.net/publication/391522222_Bridging_the_Gap_Synergizing_Anthropology_Economics_History_and_Sociology_for_Environmental_Peacebuilding

Serrat, O. (2025f). *Milestones of scholarship: From idea to contribution in environmental peacebuilding*. Unpublished manuscript, Georgetown University. https://www.researchgate.net/publication/397223335_Milestones_of_Scholarship_From_Idea_to_Contribution_in_Environmental_Peacebuilding

Serrat, O. (2025g). Preserving paradise: The Heart of Borneo Initiative. In *Anthropogenic solutions for climate change: Achieving environmental peace* (pp. 221–227). Springer.

Serrat, O. (2025h). Safeguarding Lake Chad. In *Anthropogenic solutions for climate change: Achieving environmental peace* (pp. 243–257). Springer.

Serrat, O. (2025i). Safeguarding the mother of waters: Environmental peacebuilding in the Mekong River Basin. In *Anthropogenic solutions for climate change: Achieving environmental peace* (pp. 229–239). Springer.

Swain, A., & Öjendal, J. (Eds.). (2018). *Routledge handbook of environmental conflict and peacebuilding*. Routledge.

United Nations. (2015). *Transforming our world: The 2030 Agenda for Sustainable Development* (A/RES/70/1). United Nations. https://sdgs.un.org/2030agenda

United Nations Environment Programme. (2009). *From conflict to peacebuilding: The role of natural resources and the environment*. United Nations Environment Programme. https://wedocs.unep.org/handle/20.500.11822/7867

United Nations Environment Programme. (2012). *Greening the blue helmets: Environment, natural resources and UN peacekeeping operations*. United Nations Environment Programme. https://www.unep.org/resources/report/greening-blue-helmets-environment-natural-resources-and-un-peacekeeping-operations

United Nations Secretary-General. (2010). *Progress report of the secretary-general on peacebuilding in the immediate aftermath of conflict*. United Nations.

Yin, R. (2018). *Case study research and applications: Design and methods* (6th ed.). SAGE Publications.

Zelli, F., & Krause, T. (2025). The nature of peace: Post-conflict peacebuilding and its implications for environment and livelihoods. *Ecology and Society, 30*(4), 15.

Chapter 7
Extending OECD–DAC: Designing Context-Sensitive Evaluation Frameworks for Environmental Peacebuilding

Abstract Environmental peacebuilding is maturing faster than the tools used to judge whether it works. This précis contends that evaluation must evolve in step with the field's complexity. It argues for extending the OECD–DAC's evaluation criteria so they can capture the tightly interwoven governance, ecological, and conflict dynamics that shape environmental peacebuilding. It also proposes a more adaptive, context-sensitive, and analytically robust set of supplementary evaluative criteria that address socio-ecological complexity, conflict sensitivity, legitimacy, participation, and long-term resilience. Through comparative case studies, the planned research will illuminate the causal pathways by which environmental interventions foster cooperation and contribute to "peace writ large," thereby clarifying how environmental governance supports durable peace.

7.1 Why Evaluation in Environmental Peacebuilding Needs Reimagining: Limitations of Existing Frameworks and the Case for Adaptation

Environmental peacebuilding rests on assumptions that its current evaluation tools cannot detect. It has gained prominence as ecological pressures rise, climate disruptions intensify, and resource governance becomes intertwined with questions of justice, peace, and ecological resilience. Yet the tools used to assess interventions remain anchored in conventional development evaluation frameworks that were never designed to account for the entangled dynamics of conflict and ecology. The Organisation for Economic Co-operation and Development's Development Assistance Committee (OECD–DAC) sets widely used evaluation criteria—relevance, coherence, effectiveness, efficiency, impact, and sustainability—that continue to direct practice as codified in updated guidance (OECD, 2019).[1] However,

[1] The OECD–DAC introduced a set of five evaluation criteria in 1991 to harmonize assessment practices across donors. The criteria were adopted by multilateral development agencies and incor-

O. Serrat, *Advancing Environmental Peacebuilding*, SpringerBriefs in Political Science, https://doi.org/10.1007/978-981-92-4028-9_7

these evaluation criteria register only a fraction of the non-linear and feedback-intensive interactions that characterize socio-ecological systems in fragile settings. Accordingly, evaluation often collapses causal pathways, obscuring how environmental interventions reconfigure joint action or unintentionally heighten local frictions. It also struggles to account for issues such as equity, legitimacy, and ecological resilience. A conflict-sensitive evaluative approach must therefore reinforce the peacebuilding norm of nonviolence by identifying where interventions may create incentives for coercive or violent resource competition. A conflict-sensitive evaluative approach also requires attention to the intergenerational stakes embedded in environmental governance, where today's interventions forge the ecological conditions that future communities will depend on, for better or for worse. In short, the field is evolving faster than the evaluative tools meant to guide it, leaving consequential entanglements only partially captured. This précis therefore sketches a research agenda to extend the OECD–DAC framework for environmental peacebuilding, adding context-sensitive evaluative criteria that reflect socio-ecological intricacy and conflict sensitivity while remaining usable for policymakers, practitioners, and donors. The précis argues that only by re-thinking evaluation in this way can the field move from aspirational rhetoric to evidence-based practice capable of informing policy, guiding resource allocation, and catalyzing the co-creation of enduring ecological futures.

7.2 Why Development-Oriented Evaluation Falls Short in Conflict–Ecology Settings: Missing Socio-Ecological and Conflict-Sensitive Dimensions

Environmental peacebuilding aims to transform resource-based tensions into cooperation, ecological resilience, and sustainable governance—yet its evaluative architecture remains poorly-aligned to meet these ambitions. Development-oriented evaluation tools typically assume fixed intervention boundaries, linear change pathways, and stable institutions—assumptions that are at odds with the broader evaluation logic outlined by Scriven (1991). As a result, existing evaluation practices often default to reductive models that ignore shifting baselines, sideline ecological feedback loops, and fail to surface how environmental initiatives reshape trust-building, legitimacy, or conflict trajectories. They likewise tend to overlook the culturally embedded interpretations communities bring to environmental change, cooperation, and governance. They also rarely consider how interventions affect the ecological inheritance passed to future generations, a core concern in

porated into the United Nations Evaluation Group's Norms and Standards in 2005, cementing their status as the global reference point for development evaluation. In 2019, the OECD–DAC added *coherence* as a sixth criterion (OECD, 2019) to reflect contemporary priorities such as the 2030 Agenda for Sustainable Development (United Nations, 2015).

sustainability-oriented peacebuilding. This results in obscured causal mechanisms, distorted learning, and incentives for narrow, output-focused reporting, which in turn displace attention from assessing changes in collective problem-solving or long-term system resilience.[2] Scholars and practitioners have pointed to potential evaluative domains—such as conflict sensitivity; cultural appropriateness, including recognition of Indigenous and local epistemologies; ecological resilience; legitimacy; participation; long-term environmental impact; and contributions to "peace writ large"—that existing frameworks acknowledge but do not integrate These omissions do not merely limit analytical precision; they constrain the field's ability to understand whether interventions genuinely foster shared stewardship, and they entrench structural barriers rooted in overly narrow evaluation logics that impede both practical effectiveness and scholarly progress. Only by widening the evaluative lens to match the socio-ecological complexity of the systems it seeks to assess can the field generate the kind of evidence that contributes to environmental peace—or fail to do so.

7.3 Current Landscape of Environmental Peacebuilding Scholarship

7.3.1 Intellectual and Institutional Origins of Environmental Peacebuilding

Environmental peacebuilding has coalesced into a disciplinary domain, its intellectual and institutional foundations solidified over the past five decades. The 1972 Stockholm Conference first elevated environmental protection as a global political priority, while the 1992 Rio Earth Summit linked sustainable development to peace, laying the groundwork for later policy, practice, and scholarship. Foundational contributions by Homer-Dixon (1999) and Conca and Dabelko (2002) reframed environmental stress as both a risk and an opportunity, highlighting channels through which environmental cooperation might support peacebuilding. Subsequent institutional developments—including the Environmental Peacebuilding Initiative (2005), Environmental Peacebuilding Platform (2014), and Environmental Peacebuilding Association (2018)—have given the field a clearer identity and organizational infrastructure.

[2] Because implementers expect to be judged by criteria that privilege linear outputs and short-term efficiency, the OECD–DAC framework can also unintentionally disincentivize the very environmental peacebuilding initiatives most needed in fragile socio-ecological systems.

7.3.2 Empirical Insights on Environment–Conflict Dynamics

Empirical research has established the centrality of environmental factors in conflict dynamics. Studies such as Matthew et al. (2009), which found that nearly 40% of intrastate conflicts since 1945 involved natural resources, demonstrate the importance of environmental conditions to security outcomes. This builds on the pathways for environmental cooperation outlined by Conca and Dabelko (2002), who emphasized how shared ecological challenges can create incentives for dialogue and institutional innovation. Comprehensive case-based syntheses such as Bruch et al. (2016) have demonstrated how governance and natural resource management shape post-conflict trajectories. More recent theoretical work, including Dresse et al. (2019), has articulated how environmental features can generate incentives for collaborative engagement and support peacebuilding processes. However, despite this growing empirical richness, the field still lacks a unified evaluative framework capable of translating these insights.

7.3.3 Persistent Gaps in Evaluation Practice

Existing evaluation methods remain fragmented, inconsistently applied, and often development-biased. Jean and Ernstorfer (2014) and Paffenholz (2016) underscore persistent shortcomings, including implicit and under-specified theories of change, the absence of baselines, and a tendency to overemphasize program or project outputs rather than contributions to "peace writ large." Earlier tools—such as Bush's (1998) Peace and Conflict Impact Assessment and Church and Rogers' (2006) guidance on monitoring and evaluation—opened conceptual space but remain underused and insufficiently attuned to ecological dynamics. The Environmental Law Institute's (2023) *Toolkit on Monitoring and Evaluation of Environmental Peacebuilding* acknowledges the need for adapted evaluation models but offers no operational guidance. These shortcomings reflect what Paffenholz (2016) identifies as structural fragmentation in peacebuilding evaluation, where methodological rigor lags behind conceptual ambition.

7.3.4 Policy Calls for Evidence-Driven Environmental Peacebuilding

Policy frameworks have echoed these concerns, most prominently the *White Paper on the Future of Environmental Peacebuilding* (Geneva Peacebuilding Platform et al., 2022). The White Paper calls for building an "infrastructure for peace"—the agreements, institutions, and practices that foster cooperation and resilience across ecological and social domains. It emphasizes that "monitoring, evaluation, and knowledge sharing are indispensable for demonstrating impact, ensuring accountability, and integrating environmental peacebuilding within education, funding, and governance systems" (p. 14). Earlier initiatives, such as *Greening the Blue Helmets* (United Nations Environment Programme, 2012) and the 2030 Agenda for Sustainable Development (United Nations, 2015) had already substantiated the institutional relevance of evaluative rigor. Yet despite these calls, the field still lacks a coherent evaluative architecture that can align normative aspirations with practical assessment tools. This gap affirms the need for a context-attuned framework that makes visible the hybrid governance–ecology–conflict dynamics at the heart of environmental peacebuilding.

7.3.5 Positioning the Planned Research within the Field

Responding to the White Paper's call for an "infrastructure for peace research," the planned research will position itself as a practical and conceptual bridge between environmental peacebuilding theory and the evaluative tools needed to assess it. The field has developed robust insights into cooperation, governance, and ecological interdependence, yet it still lacks an assessment schema capable of capturing how these dynamics unfold in practice. The research will close this gap by developing three mutually reinforcing interventions.

First, the research will ground the principles of cooperation, sustainability, and inclusive governance within existing policy and scholarly debates, clarifying their analytical relevance rather than treating them as purely normative aspirations. Second, it will extend the OECD–DAC's evaluation criteria by integrating a set of candidate evaluative criteria identified in the literature—such as adaptability and resilience, conflict sensitivity and "do no harm", cultural sensitivity, inclusivity and participation, legitimacy, long-term environmental impact, and "peace writ large"—thereby recalibrating the framework to the enmeshed governance–ecology–conflict realities that define the field. These domains will be presented as provisional rather than definitive, though existing scholarship suggests they are likely to remain central; their final form will be shaped through empirical analysis. Third, the research will apply these adapted criteria through comparative case studies, tracing causal pathways and identifying mechanisms that show how environmental interventions might contribute to peace.

Together, these strands of inquiry confront the field's central challenge: the absence of an evaluative framework that is both theoretically grounded and operationally usable across diverse contexts. By drawing on insights from Lederach (1997), Pelling (2011), and Platteau (2004), the research will ensure that the proposed criteria remain attentive to social legitimacy, ecological sustainability, and governance risks—including elite capture and uneven participation. Moreover, by designing the framework to be modular rather than prescriptive, the research will address a common critique that environmental peacebuilding is too context-specific for standardized evaluation. Instead, it will show how a flexible architecture can maintain comparability while remaining responsive to local socio-ecological realities.

7.3.6 Philosophical and Conceptual Foundations of the Research Design

The research design will be grounded in a set of philosophical commitments that together provide a coherent foundation for developing context-attuned evaluative criteria for environmental peacebuilding. The ontology of critical realism will anchor the inquiry by emphasizing underlying causal mechanisms—a philosophy of science well-suited to socio-ecological systems characterized by layered structures, emergent properties, and non-linear interactions. This orientation will enable the research to move beyond surface-level correlations and examine how environmental governance, institutional arrangements, and conflict dynamics interact, co-evolve, and generate observable outcomes. The epistemology of social constructivism will extend this orientation by highlighting how actors interpret collaborative engagement, environmental change, and governance, showing that meaning-making processes—rooted in culturally grounded worldviews—shape perceptions of legitimacy and stewardship and, in turn, institutional behavior. The methodology of pragmatism will complement these foundations by prioritizing methodological choices that yield actionable insights for policymakers and practitioners, consistent with developmental evaluation approaches outlined by Patton and Campbell-Patton (2021). Figure 7.1 illustrates how these philosophical commitments provide an integrated foundation for the research, justifying the use of comparative case studies,

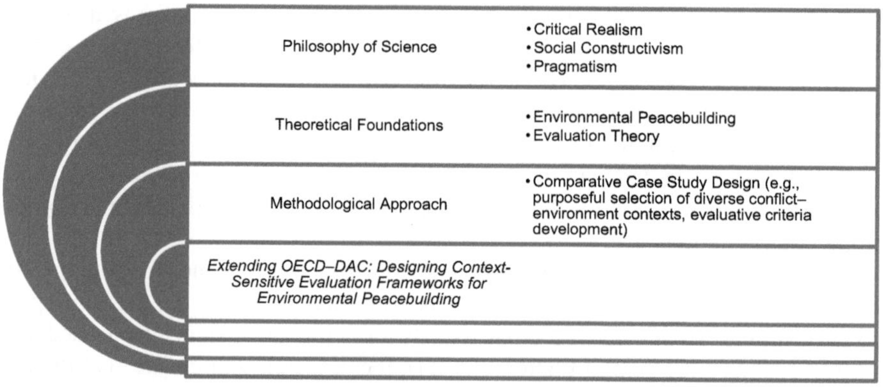

Fig. 7.1 Conceptual Framework for the Research. Note. This figure sets out the conceptual framework that will guide the research. It specifies that the epistemological commitments of critical realism, social constructivism, and pragmatism will inform the development of context-sensitive evaluative criteria and support the use of comparative case studies to examine causal mechanisms in environmental peacebuilding

supporting the development of hybrid evaluative criteria, and guiding the analysis toward uncovering mechanisms rather than documenting patterns.

7.4 Research Design and Analytical Strategy

7.4.1 The Research Question

The research will center on a core question: How can evaluative criteria be tailored to capture the conflict-sensitive, ecological, and governance dimensions of environmental peacebuilding across diverse contexts? To pursue this, three sub-questions will guide the inquiry:

- What limitations of the OECD–DAC framework constrain its applicability to environmental peacebuilding, and how might these limitations be addressed through adaptation or extension?
- Which supplementary evaluative criteria are necessary to assess environmental peacebuilding interventions more comprehensively?
- How do selected case studies illustrate the contextual relevance, feasibility, and explanatory power of the proposed hybrid evaluative criteria?

Together, these questions build squarely on the evaluative gaps outlined earlier and structure the development of a more context-responsive framework.

7.4.2 Selecting Cases that Reflect Global Variation

Environmental peacebuilding outcomes emerge from locally shaped interactions that resist isolation through experimental or purely quantitative methods. Therefore, a comparative case study design will structure the research, drawing on qualitative methods well suited to the complex interplay between ecological systems and socio-political dynamics. This design strikes a deliberate balance between comparative reach and qualitative depth, enabling the research to trace real-world complexity, triangulate evidence, and generate nuanced theoretical insights.

The research will purposefully select four to five cases that vary in conflict patterns, ecological conditions, and governance arrangements. Illustrative cases, grouped by ecological system, include:

- The Amazon Basin (South America), where resource governance intersects with Indigenous rights and transboundary cooperation.
- The Heart of Borneo Initiative (Southeast Asia), where biodiversity conservation unfolds amid competing national development agendas.
- The Indus Waters Treaty (South Asia), where a long-standing institutional arrangement balances hydropolitics with conflict sensitivity.

- The Mekong River (East and Southeast Asia), where contested development highlights ecological interdependence across multiple states.
- The Aral Sea (Central Asia), where large-scale water diversion has produced severe ecological collapse with lasting socio-economic and regional security implications.
- Lake Chad (Africa), where ecological degradation converges with insecurity, creating complex environmental peacebuilding challenges.
- The Caribbean, where climate adaptation strategies are explicitly framed through conflict-sensitive lenses.

These cases span diverse ecological zones, governance regimes, and conflict intensities, allowing for comparative insights into how principles of cooperation, sustainability, and inclusive governance are either enacted—or overlooked—in practice. The prevalence of transboundary ecosystems across these cases further strengthens the design, as such settings illuminate how institutions, ecological interdependence, and conflict dynamics co-evolve.

7.4.3 Data Sources and Analytical Techniques

Methodologically, the research will rely primarily on document-based qualitative methods, supplemented—where feasible—by practitioner conversations:

- **Document analysis** of treaties, institutional reports, NGO evaluations, and donor frameworks to trace evaluative criteria in use.
- **Practitioner conversations**, where feasible, will serve as supplementary interpretive inputs rather than sources of systematically comparable data.[3]
- **Process tracing** to uncover the causal pathways that link principles to outcomes.
- **Cross-case coding** of evaluative criteria to identify patterns, divergences, and innovations beyond OECD–DAC.

The analysis will synthesize inductive pattern-finding with abductive inference to identify the most plausible causal mechanisms connecting interventions to outcomes. Together, these methods will ensure that the research can trace both context-specific variation and cross-case regularities, strengthening the explanatory power of the proposed evaluative framework.

[3] Given the wide variations across cases—geography, ecological systems, governance arrangements, historical periods, institutional continuity, and actor turnover—it would be neither feasible nor methodologically sound to expect a consistent set of interviews. Any practitioner conversations that do occur will therefore serve only as supplementary interpretive inputs rather than comparable data. If interviews are not possible, the analysis will rely primarily on documentary and secondary sources.

7.4.4 Study Outline

The research will unfold through the following structure, moving from foundational concepts to applied analysis:

- **Introduction**—Statement of problem, research question (and subquestions), and significance.
- **Literature Review**—Foundations of environmental peacebuilding; methodological gaps in evaluation.
- **Analytical Framework**—Principles, criteria, and pathways; adaptation of OECD–DAC; integration of values.
- **Methodology**—Comparative case study design, data collection methods, analytical strategies.
- **Case Studies**—Detailed analysis of each selected case.
- **Cross-Case Analysis**—Comparative synthesis, highlighting patterns and lessons learned.
- **Discussion**—Implications for theory, policy, and practice.
- **Conclusion**—Contributions to evaluation; recommendations for future research and practice.

The cross-case analysis will feed back into the refinement of the supplementary evaluative criteria, allowing the framework to evolve in response to empirical variation across ecological and governance contexts. This iterative structure will reinforce the research's exploratory orientation, ensuring that criteria are refined through evidence rather than predetermined assumptions.

7.4.5 Research Schedule

Table 7.1 lays out the timetable for the research, mapping the progression from the early literature review through data collection, analysis, drafting, and final revisions. Each phase will build on the last, creating a coherent and manageable workflow. The schedule is designed to be flexible, allowing the research design to adapt to data availability while preserving the integrity of the comparative structure.

7.5 Why This Research Matters: Operational and Theoretical Contributions

Environmental peacebuilding will continue to expand in scope and ambition as ecological pressures intensify and governance challenges deepen, yet its evaluation tools will remain constrained if they continue to rely on frameworks designed for more stable, linear development contexts. Current evaluative practice struggles to

Table 7.1 Timetable for the Research

Months	Activity
1–3	Conduct literature review and refine the analytical framework.
4–6	Undertake preliminary case scoping, finalize case selection, and design research strategy.
7–12	Collect data (documents, datasets, ecological monitoring, and, where feasible, practitioner conversations), organized around sub-questions, recognizing that access and feasibility may vary across cases.
13–15	Carry out comparative case analysis, with flexibility to adjust if case access proves challenging.
16–18	Draft core chapters (Introduction, Literature Review, Analytical Framework, Methodology).
19–20	Draft remaining chapters (Case Studies, Cross-Case Analysis, Discussion, Conclusion).
21–23	Revise full manuscript and submit for review.

Note. This timetable may shift depending on data availability. If access to a selected case becomes constrained, the research design allows for substituting comparable cases or adjusting the depth of analysis without compromising the overall comparative structure

register the intertwined governance, ecological, and conflict dynamics that shape outcomes in fragile socio-ecological systems. The planned research will fill this gap by developing a context-sensitive evaluative framework that extends the OECD–DAC's evaluation criteria within a broader architecture grounded in core principles and supplementary evaluative criteria tailored to environmental peacebuilding. Rather than treating the OECD–DAC framework as the foundation, the research will position it as one component within a hybrid evaluative architecture designed to capture socio-ecological complexity, conflict sensitivity, and long-term governance dynamics.

The practical contribution of the research will be to translate core principles—cooperation, sustainability, and inclusive governance—into a structured suite of supplementary evaluative criteria that can guide real-world decision-making. By integrating domains such as adaptability and resilience, conflict sensitivity and "do no harm", cultural sensitivity, inclusivity and participation, legitimacy, long-term environmental impact, and "peace writ large", the framework will clarify the causal pathways through which environmental interventions shape peacebuilding trajectories. This will also help identify where interventions may inadvertently reinforce inequities, heighten tensions, or undermine ecological foundations. In doing so, the research will produce a more context-sensitive and analytically robust set of tools for learning, accountability, and strategic planning.

The theoretical contribution will be to link environmental peacebuilding and evaluation in a way that reflects the hybrid systems under study. Grounded in critical realism, social constructivism, and pragmatism, the research will treat cooperation, sustainability, and governance not as abstract aspirations but as mechanisms that can be empirically traced across diverse contexts. This orientation will clarify how environmental governance arrangements, institutional choices, and socio-ecological feedbacks interact to produce observable outcomes. It will also show how the

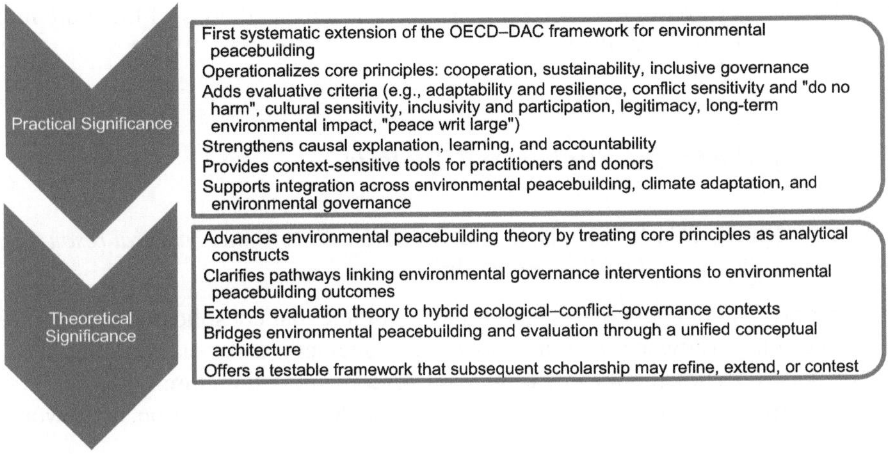

Fig. 7.2 Significance of the Research

OECD–DAC criteria—long the global reference point for evaluation—can be extended and complemented by supplementary evaluative criteria to better reflect ecological–conflict settings.

Figure 7.2 points up the practical and theoretical significance of the research. Together, these contributions will position the research as a step toward an evaluative architecture that better matches the field's conceptual maturity. By offering a modular, context-sensitive framework, the research will enable more rigorous evidence generation, strengthen institutional learning, and align environmental peacebuilding practice with the socio-ecological realities of the Anthropocene. The resulting evaluative approach will be capable of informing policy, guiding resource allocation, and supporting the co-creation of more peaceful and ecologically resilient futures.

7.6 Annotated Bibliography

The annotated bibliography below reflects a rich landscape of environmental peacebuilding scholarship: foundational works and global case studies offer strong conceptual grounding and practical insights, while methodological guides and policy frameworks provide useful tools and institutional direction. Across these sources, however, a consistent limitation stands out: many remain theoretically abstract, development-biased, or insufficiently operationalized, offering little in the way of concrete evaluative criteria for assessing environmental peacebuilding initiatives. The research sketched out in this précis will address this gap by integrating these diverse contributions into a coherent, context-sensitive evaluative framework that integrates conflict sensitivity, ecological resilience, governance inclusivity, and social legitimacy. In doing so, it will transform a fragmented body of literature into a coherent evaluative architecture.

7.6.1 Applied Case Studies and Toolkits

Bruch, C., Muffett, C., & Nichols, S. (Eds.). (2016). *Governance, natural resources, and post-conflict peacebuilding*. Routledge.

- **Strength:** A comprehensive, authoritative volume that consolidated global case studies on governance, natural resource management, and peacebuilding, making it a cornerstone in the environmental peacebuilding literature.
- **Limitation:** Provided limited operational guidance on evaluation frameworks and metrics.
- **The Planned Research's Contribution:** The research will build on Bruch et al.'s governance and resource insights by embedding structured evaluative criteria, ensuring that environmental peacebuilding interventions are assessed for both ecological resilience and social legitimacy.

Environmental Law Institute. (2023). *Toolkit on monitoring and evaluation for environmental peacebuilding*. Environmental Law Institute.

- **Strength:** Offered practical tools for evaluative thinking in environmental peacebuilding.
- **Limitation:** Toolkit was broad, not deeply theorized.
- **The Planned Research's Contribution:** The research will integrate toolkit insights into a structured evaluative framework, moving beyond broad guidance to concrete operational tools.

7.6.2 Foundational Theory and Conceptual Works

Conca, K., & Dabelko, G. (Eds.). (2002). *Environmental peacemaking*. Johns Hopkins University Press.

- **Strength:** Offered a foundational text linking environment and peace with diverse case studies.
- **Limitation:** Gave limited attention to evaluation frameworks.
- **The Planned Research's Contribution:** The research will operationalize Conca and Dabelko's conceptual insights through evaluative criteria, offering the first structured evaluative lens that links environment and peace.

Dresse, A., Fischhendler, I., Nielsen, J., & Wolf, A. (2019). *Environmental peacebuilding: Towards a theoretical framework. Global Environmental Change, 58*, 101,956.

- **Strength:** Advanced theoretical foundations of environmental peacebuilding.
- **Limitation:** Framework remained abstract, lacking operational evaluation.
- **The Planned Research's Contribution:** The research will provide evaluative architecture to test and refine theory, ensuring ecological and conflict-sensitive dimensions are explicitly measured.

Homer-Dixon, T. (1999). *Environment, scarcity, and violence*. Princeton University Press.

- **Strength:** Linked resource scarcity to conflict.
- **Limitation:** Focused on scarcity, less on cooperation or evaluation.
- **The Planned Research's Contribution:** The research will move beyond scarcity to operationalize peacebuilding through evaluative criteria that emphasize cooperation and resilience.

Lederach, J. (1997). *Building peace: Sustainable reconciliation in divided societies*. United States Institute of Peace Press.

- **Strength:** Presented seminal work on reconciliation and legitimacy in peacebuilding.
- **Limitation:** Focused primarily on social reconciliation, less on ecological dimensions.
- **The Planned Research's Contribution:** The research will broaden reconciliation frameworks by integrating ecological sustainability and evaluative criteria, bridging social legitimacy with environmental resilience.

Pelling, M. (2011). *Adaptation to climate change: From resilience to transformation*. Routledge.

- **Strength:** Provided evaluative insights into resilience and adaptation.
- **Limitation:** Focused on climate adaptation, not directly on peacebuilding.
- **The Planned Research's Contribution:** The research will integrate resilience thinking into supplementary evaluative criteria, connecting climate adaptation insights to peacebuilding evaluation.

Platteau, J.-P. (2004). *Monitoring elite capture in community-driven development. Development and Change*, *35*(2), 223–246.

- **Strength:** Highlighted risks of elite capture in participatory governance.
- **Limitation:** Focused on development, not peacebuilding.
- **The Planned Research's Contribution:** The research will integrate elite capture concerns into inclusivity criteria, ensuring evaluations account for governance legitimacy.

7.6.3 Methodological Guidance and Evaluation Frameworks

Bush, K. (1998). *A measure of peace: Peace and conflict impact assessment (PCIA) of development projects*. International Development Research Centre.

- **Strength:** Introduced PCIA, pioneering evaluative thinking.
- **Limitation:** Was under-tested empirically; limited ecological dimension.
- **The Planned Research's Contribution:** The research will adapt Bush's evaluative lens to hybrid ecological–peacebuilding contexts, filling the gap in empirical application.

Church, C., & Rogers, M. (2006). *Designing for results: Integrating monitoring and evaluation in conflict transformation programs*. Search for Common Ground.

- **Strength:** Provided a practical guide for integrating monitoring and evaluation in conflict transformation.
- **Limitation:** Focused on conflict programs, not ecological dimensions.
- **The Planned Research's Contribution:** The research will extend Church and Rogers's evaluative strategies to the field of environmental peacebuilding, embedding ecological resilience into conflict-sensitive evaluation.

Jean, I., & Ernstorfer, A. (2014). *Conflict-sensitive monitoring and evaluation: Guidance note.* CDA Collaborative Learning Projects.

- **Strength:** Provided practical guidance on adapting monitoring and evaluation to conflict-affected contexts, emphasizing "peace writ large."
- **Limitation:** Focused on conflict sensitivity but did not integrate ecological dimensions.
- **The Planned Research's Contribution:** The research will enrich Jean and Ernstorfer's conflict-sensitive approach by embedding ecological resilience and sustainability criteria.

Organization for Economic Co-operation and Development. (2019). *Better criteria for better evaluation: Revised evaluation criteria defined.* OECD Publishing.

- **Strength:** Established global evaluation standards, widely adopted.
- **Limitation:** Was development-centric, not tailored to peacebuilding.
- **The Planned Research's Contribution:** The research will provide the first systematic attempt to adapt and extend the OECD–DAC framework with ecological and conflict-sensitive criteria.

Paffenholz, T. (2016). Making peace stick: Lessons from peacebuilding evaluation. *Journal of Peacebuilding & Development*, *11*(1), 1–17.

- **Strength:** Critiqued weak theories of change and the absence of baselines, highlighting how these undermine causal claims in peacebuilding evaluation.
- **Limitation:** Focused narrowly on theories of change and baselines, without ecological resilience.

- **The Planned Research's Contribution:** The research will adapt Paffenholz's evaluative insights to environmental contexts, strengthening causal claims with ecological baselines.

Patton, M., & Campbell-Patton, C. (2021). *Utilization-focused evaluation* (5th ed.). Sage Publications.

- **Strength:** Offered a practical approach to evaluation that prioritizes use, decision relevance, and organizational learning.
- **Limitation:** Was not tailored to ecological or conflict-affected contexts.
- **The Planned Research's Contribution:** The research will apply Patton's utilization-focused approach to environmental peacebuilding, ensuring evaluative criteria are not only rigorous but also practically useful and adaptable for policymakers, practitioners, and scholars.

Scriven, M. (1991). *Evaluation thesaurus* (4th ed.). Sage Publications.

- **Strength:** Established foundational concepts of evaluation as truth-seeking, improvement, and accountability.
- **Limitation:** Was highly conceptual, offering limited operational tools for application in complex environmental-peacebuilding contexts.
- **The Planned Research's Contribution:** The research will propose context-sensitive pathways to operationalize Scriven's evaluative principles, acknowledging institutional and ecological constraints.

7.6.4 Policy Frameworks and Institutional Reports

Geneva Peacebuilding Platform, Environmental Peacebuilding Association, & PeaceNexus Foundation. (2022). *White paper on the future of environmental peacebuilding*. Geneva Peacebuilding Platform.

- **Strength:** Identified institutional priorities for the field.
- **Limitation:** Lacked methodological depth.
- **The Planned Research's Contribution:** The research will seek to embed evaluative rigor into institutional commitments, offering practical pathways for uptake.

Matthew, R., Brown, O., & Jensen, D. (2009). *From conflict to peacebuilding: The role of natural resources and the environment*. United Nations Environment Programme.

- **Strength:** Linked natural resources to peacebuilding.
- **Limitation:** Was a broad survey, with limited methodological depth.
- **The Planned Research's Contribution:** The research will refine Matthew et al.'s insights with tailored evaluative criteria, bridging broad surveys with operational tools.

United Nations. (2015). *Transforming our world: The 2030 agenda for sustainable development*. United Nations.

- **Strength:** Established global goals linking peace, justice, and climate action.
- **Limitation:** Delineated a broad agenda, not tailored to environmental peacebuilding.
- **The Planned Research's Contribution:** The research will situate environmental peacebuilding within Sustainable Development Goals (SDGs) 13 and 16, demonstrating how evaluation can connect peace, justice, and climate action.

United Nations Environment Programme. (2012). *Greening the blue helmets: Environment, natural resources, and UN peacekeeping operations*. UNEP.

- **Strength:** Linked natural resources to peacekeeping.
- **Limitation:** Focused on peacekeeping, not broader peacebuilding.
- **The Planned Research's Contribution:** The research will bridge preventive diplomacy insights into proactive environmental peacebuilding, foregrounding ecological dimensions that are often overlooked.

7.7 Future Directions: Reorienting Evaluation from Anthropocentrism to Ecocentrism

Environmental peacebuilding will continue to evolve as ecological pressures intensify and governance innovations emerge, and its evaluation frameworks will need to adapt in parallel. A promising line of inquiry for future research concerns how evaluative criteria might engage with ecocentric governance models, including rights-of-nature approaches that treat ecosystems as rights-bearing entities rather than solely as resources for human use. This perspective is evident in Indigenous governance systems and in jurisdictions experimenting with ecological jurisprudence, and it raises far-reaching questions about how evaluation can register ecological integrity as an outcome in its own right.

While this précis has focused on extending the OECD–DAC framework and proposing supplementary evaluative criteria to account for conflict-sensitive, socio-ecological, and governance dynamics, the rising normative significance of rights-of-nature highlights a further frontier: developing evaluative tools capable of operating in contexts where legitimacy, collaborative engagement, and long-term resilience are interpreted through relational or ecocentric worldviews. Such contexts require evaluative approaches that can account for reciprocal human–ecosystem relationships, non-human agency, and long-term ecological flourishing as central components of peacebuilding success.

Exploring how ecocentric perspectives intersect with environmental peacebuilding would deepen the field's conceptual foundations and strengthen its ability to assess initiatives in settings where human and ecological well-being are understood as mutually constitutive rather than hierarchically ordered. If environmental peacebuilding is to meet the demands of a rapidly changing world, its evaluative tools must be capable not only of measuring cooperation and governance performance, but also of recognizing when ecosystems themselves become drivers of peace and when ecological integrity functions as a primary peacebuilding outcome rather than a secondary benefit.

Chapter 8
The Tonle Sap Initiative as Proto-Environmental Peacebuilding

Abstract The Tonle Sap Initiative, launched by the Asian Development Bank in the early 2000s, offers an overlooked case of proto-environmental peacebuilding. Developed in Cambodia's fragile post-conflict context, it combined participatory resource management, multi-level coordination, and shared knowledge systems to address structural vulnerabilities in the Tonle Sap basin. Although not framed as peacebuilding, it operationalized mechanisms later formalized in environmental peacebuilding theory and shows that development practice was generating peace-enabling conditions before the field had a name. It illustrates how integrated environmental governance can function as peace infrastructure and broadens the field's understanding of where environmental peacebuilding emerges and how it takes shape.

8.1 The Tonle Sap Initiative in Environmental Peacebuilding

Environmental peacebuilding has emerged as a valuable framework for understanding how cooperation around natural resources can contribute to stability and resilience. The field has grown rapidly, with scholars examining how shared environmental challenges can foster cooperation, reduce tensions, and strengthen institutions. Much of this literature traces the field's origins to early-2000s academic debates, particularly the publication of *Environmental Peacemaking* (Conca & Dabelko, 2002), which drew on six regionally grounded case studies—from the Baltic Sea to the U.S.–Mexico border—to illustrate how environmental cooperation

I launched the Tonle Sap Initiative in 2001 while serving as a senior project economist (natural resources) at the Asian Development Bank. I led the program until 2006. The views expressed in this précis are mine and do not necessarily reflect the views or policies of the Asian Development Bank, or its Board of Governors, or the governments they represent.

O. Serrat, *Advancing Environmental Peacebuilding*, SpringerBriefs in Political Science, https://doi.org/10.1007/978-981-92-4028-9_8

could support broader social and political goals.[1] This précis argues that the Tonle Sap Initiative (TSI), conceptualized by the Asian Development Bank in 2001, represents an analytically significant parallel example that demonstrated environmental peacebuilding in practice.

Future Solutions Now: The Tonle Sap Initiative (Serrat, 2002, 2003a, 2003b, 2003c, 2004a, 2004b, 2005, 2006), a six-monthly, practice-oriented knowledge series, showcased ADB's assistance to Cambodia under the TSI. Between 2002 and 2006, this knowledge and learning platform documented ongoing work, distilled lessons learned, and offered a running commentary on development practice, serving not merely as a documentation platform but as an incubator for integrated governance thinking that prefigured environmental peacebuilding principles with concrete, operational deliverables. It addressed a wide range of themes including the Tonle Sap in ADB's country partnership strategy; the Tonle Sap Environmental Management Project (2002); the Tonle Sap Biosphere Reserve Environmental Information Database[2]; the value of wetlands; poverty dynamics around the lake; drivers of wetland loss, inland fisheries research and development; biosphere reserve management; the biodiversity of the Tonle Sap; national environmental education and awareness; community organization for natural resource management; the formulation of the *Tonle Sap Basin Strategy* (Serrat et al., 2005); the Tonle Sap Sustainable Livelihoods Project (2005); Tonle Sap fisheries; and approaches to reconciling multiple demands through basin management organizations.

Although the TSI was not framed as a peacebuilding initiative—positioning itself instead as a development and environmental governance program—its design, implementation modalities, and institutional architecture anticipated with remarkable precision the mechanisms that scholars would later identify as central to environmental peacebuilding.

The TSI's significance lies not only in its outcomes but in its timing. It reveals that the conceptual foundations of environmental peacebuilding were being

[1] The case studies examined in the volume included post–Cold War collaboration around the Baltic Sea (late 1980s–early 2000s); transboundary water-management initiatives in South Asia, particularly within the Ganges–Brahmaputra–Meghna basin (mid-1970s–early 2000s); regional coordination to address the shrinking Aral Sea in Central Asia (late 1980s–early 2000s); post-apartheid natural-resource cooperation frameworks in southern Africa (1994–early 2000s); joint environmental monitoring and fisheries management in the Caspian Sea (early 1990s–early 2000s); and cross-border conservation and pollution-control mechanisms along the U.S.–Mexico border (1983–early 2000s). Conca and Dabelko's (2002) causal claims, methodological limitations, and underdeveloped treatment of power and conflict dynamics have been critiqued. Even so, these critiques helped steer subsequent research toward more critical, empirically grounded, and justice-oriented approaches.

[2] The Tonle Sap Biosphere Reserve was designated in 2001 under the United Nations Educational, Scientific, and Cultural Organization's Man and the Biosphere Programme, which provides an international framework for integrating biodiversity conservation, sustainable development, and scientific research. The Tonle Sap Biosphere Reserve Secretariat serves as Cambodia's coordinating body for this designation, facilitating cooperation among ministries, provincial authorities, research institutions, and local communities; supporting ecological monitoring and research; and preparing periodic reports.

developed in practice at the same time scholars were formalizing the field in the early 2000s, with full institutionalization occurring in the 2010s through dedicated research networks, peer-reviewed journals, and the integration of environmental peacebuilding principles into global policy frameworks (Serrat, 2026). This insight expands the field's intellectual history and highlights the contributions of development institutions and practitioners (Serrat, 2026).

8.2 Post-Conflict Cambodia and the Tonle Sap Basin

In the early 2000s, Cambodia remained in a fragile post-conflict state. The country was emerging from decades of civil war, political instability, and institutional fragmentation. Governance capacity was uneven, particularly in rural areas, and state institutions struggled to regulate resource use, enforce environmental laws, or provide basic services. Poverty remained high—around 35–40% nationally, and higher in rural areas—leaving households heavily dependent on natural resources for subsistence. These conditions created precisely the kinds of environmental–institutional vulnerabilities that environmental security scholars warn can escalate into conflict.

The demographic landscape added further strain. Following the population collapse during the Khmer Rouge period (1975–1979), Cambodia experienced rapid demographic expansion throughout the 1980s and 1990s. By the early 2000s, the population had reached roughly 12–13 million, characterized by a large youth cohort and limited livelihood opportunities. Many households migrated toward resource-rich areas—particularly the Tonle Sap region—in search of fisheries, forest products, and arable land, intensifying pressure on ecosystems that were already under significant stress.

In the Tonle Sap basin, the combined effects of population growth, resettlement, and weak regulatory oversight accelerated natural-resource extraction. Flooded forests—once covering over 1 million ha—were shrinking rapidly due to fuelwood collection and agricultural expansion. Wetlands were converted for cultivation or settlement, and fishing pressure increased sharply, often through illegal or destructive gear. These trends reflected both post-conflict recovery dynamics and the absence of regulatory mechanisms capable of balancing immediate livelihood needs with long-term ecological sustainability. In short, the Tonle Sap region exemplified worsening environmental scarcity interacting with weak institutions—conditions later theorized as drivers of social tension.

These pressures were compounded by the institutional legacy of the fishing-lot system and the slow emergence of community-based management. Large commercial lots controlled access to the most productive fishing grounds, leaving small-scale fishers with little legal space to operate. As populations grew and catches declined, encroachment into lot areas increased, prompting coercive enforcement by lot holders and recurrent low-intensity violence as armed guards seized gear, rammed boats, and confronted fishers entering restricted waters. The 2000–2001 reforms released 56% of lot areas to the public, but the newly created community fisheries lacked clear mandates and resources, producing a governance vacuum that intensified competition, weakened ecological and social resilience, and heightened the risk of localized conflict—making cooperative governance particularly urgent.

8.3 The Tonle Sap as a Socio-Ecological System

The Tonle Sap is the ecological heart of Cambodia. It is Southeast Asia's largest freshwater body and one of the world's most productive inland fisheries. Its hydrological cycle is unique: during the monsoon, the Mekong River reverses the flow of the Tonle Sap River, expanding the lake from roughly 1040 mile^2 (about 2700 km^2) to more than 6180 mile^2 (about 16,000 km^2).[3] This annual flood pulse drives an exceptionally productive ecosystem that sustains millions of people and has done so for centuries; the stability and abundance of this system were central to the rise and endurance of the Khmer civilization (c. ninth–fifteenth century), which drew upon the extraordinary wealth generated by the Tonle Sap's floodplain to build monumental complexes such as Angkor Wat, Bayon, and Ta Prohm and sustain its political power. This deep historical dependence underscores why governance failures in the basin carry ecological, political, and social consequences.

By the early 2000s, 4.5 million people—about one-third of Cambodia's population—depended directly or indirectly on the Tonle Sap basin's freshwater resources. Between 1.2 and 1.5 million relied directly on fishing and fish-related activities. The lake produced 400,000–500,000 tons of fish annually, accounting for around 60% of Cambodia's inland catch and supplying up to 75% of national animal protein intake. Aquatic vegetation, flooded forests, and wetlands provided critical breeding and nursery grounds, while the receding floodwaters concentrated fish in accessible areas, sustaining both subsistence and commercial fisheries. Beyond fisheries, the lake's floodplain moderated hydrological extremes, supported regionally significant biodiversity, and provided habitat for numerous endangered species. These ecological services stabilized livelihoods, shaped cultural practices, and underpinned food security across the basin.

Yet the system's productivity was highly sensitive to disturbance. It depended on the integrity of the flood pulse, the health of the flooded forests, and governance arrangements capable of managing resource use equitably. Disruptions—from

[3] Seasonal inundation of the Tonle Sap floodplain replenishes soils with alluvial nutrients and expands aquatic habitats, enabling intensive rice cultivation and exceptionally high fish biomass. The annual flood pulse also triggers large-scale fish migrations: many species move from the Mekong into the lake to spawn in the flooded forests, where nutrient-rich waters and dense vegetation provide ideal nursery grounds. These inundated forests function as critical feeding and growth habitats, supporting rapid juvenile development. As the waters recede, young fish concentrate in channels and the main lake, producing seasonally dense fish assemblages that underpin a level of productivity rivaled globally only by the Amazon floodplain and the African Great Lakes, though each operates through distinct ecological and hydrological dynamics.

upstream hydropower development to local overexploitation—had cascading effects on both ecosystems and communities. Recognizing this socio-ecological interdependence was essential for designing governance strategies capable of sustaining the Tonle Sap's functions and supporting long-term stability, a premise the TSI adopted by grounding its interventions in a basin-wide systems understanding.

The *Tonle Sap Basin Strategy* (Serrat et al., 2005) identified several interlinked biophysical, governance, and socioeconomic pressures: rapid population growth and settlement expansion; intensifying exploitation of natural resources, especially fisheries; agricultural expansion and land conversion; loss and degradation of flooded forests; hydrological and ecological stress; weak regulatory capacity; limited coordination among ministries; and rising livelihood insecurity. These pressures had direct implications for social cohesion, economic stability, and state legitimacy—conditions central to peacebuilding.

8.4 Environmental Governance as Peace Infrastructure

Although the term "environmental peacebuilding" was not yet in use, the conditions in the Tonle Sap region reflected the dynamics that environmental security scholars had begun to identify. Homer-Dixon (1999) argued that environmental scarcity, household economic insecurity, and weak institutions and can interact to produce social tension and conflict. The Tonle Sap region exhibited precisely this combination of vulnerabilities. The *Tonle Sap Basin Strategy* (Serrat et al., 2005) recognized that addressing these challenges required more than technical interventions. It required cross-sectoral coordination, multi-level oversight, participatory planning, institutional strengthening, livelihood diversification, and transparent resource management. These elements would later become core pillars of environmental peacebuilding theory, underscoring the TSI's anticipatory character. The TSI thus emerged in a context where environmental governance was inseparable from peacebuilding, even if not explicitly labeled as such.

8.5 Designing the Tonle Sap Initiative

8.5.1 From Strategic Vision to Operational Framework

Framed early by the *Future Solutions Now* knowledge series, the *Tonle Sap Basin Strategy* (Serrat et al., 2005) appeared as a formal strategic document in support of ADB's Country Strategy and Program (2005–2007) for Cambodia. Yet the strategic thinking that underpinned it had already been evolving for several years. From 2002, ADB used the *Future Solutions Now* knowledge series to explore a basin-wide approach to environmental governance, test emerging ideas across departments, and cultivate internal acceptance for a more integrated framework. These early representations show that the TSI's conceptual foundations were well developed before the strategy's official release.[4] This sequencing is crucial: it reveals the TSI as a governance experiment aligned with evolving thinking in environmental cooperation.

By the time the *Tonle Sap Basin Strategy* (Serrat et al., 2005) articulated Cambodia's development challenges as inseparable from the condition of its natural resource base—highlighting how environmental degradation and livelihood vulnerability reinforced one another—the institution had already spent years refining this diagnosis in consultation with a broad spectrum of institutional actors, including central government ministries, sub-national administrations, nongovernment organizations, community representatives, national and international research institutes, and development partners engaged in the Tonle Sap basin. The published strategy therefore did not inaugurate a new direction so much as codify strategic thinking that had matured through earlier internal and external work, positioning environmental governance as essential not only for development but for long-term stability. This implicit framing mirrors the preventive logic that environmental peacebuilding scholars would later formalize.

Figure 8.1 below illustrates the strategic principles, crosscutting themes, and operating principles that guided the TSI's regular activities and annual operating outputs toward its overarching goal of poverty reduction. These principles shaped work across the TSI's two core intervention areas—(1) rural development and the environment and (2) human development—and supported progress toward three development objectives: (1) pro-poor, sustainable economic growth, (2) access to assets, and (3) natural-resource and environmental management.

8.5.2 Multi-Level Governance Structures

One of the TSI's most significant innovations was its multi-level governance architecture. The initiative established coordination mechanisms across national ministries, provincial authorities, the Tonle Sap Biosphere Reserve Secretariat (initially housed in the Ministry of Environment), community-based organizations, and international partners. This architecture was elaborated in *From Strategy to Practice: The Tonle Sap Initiative* (Serrat & Moffatt, 2006), which emphasized the importance of institutional innovation in tackling the region's socio-ecological challenges. By reducing fragmentation and creating predictable channels of interaction, the TSI built the institutional scaffolding for cooperative environmental governance—an essential peacebuilding function.

[4] The sector study mission that drafted the *Tonle Sap Basin Strategy* (Serrat et al., 2005) was fielded in mid-2003.

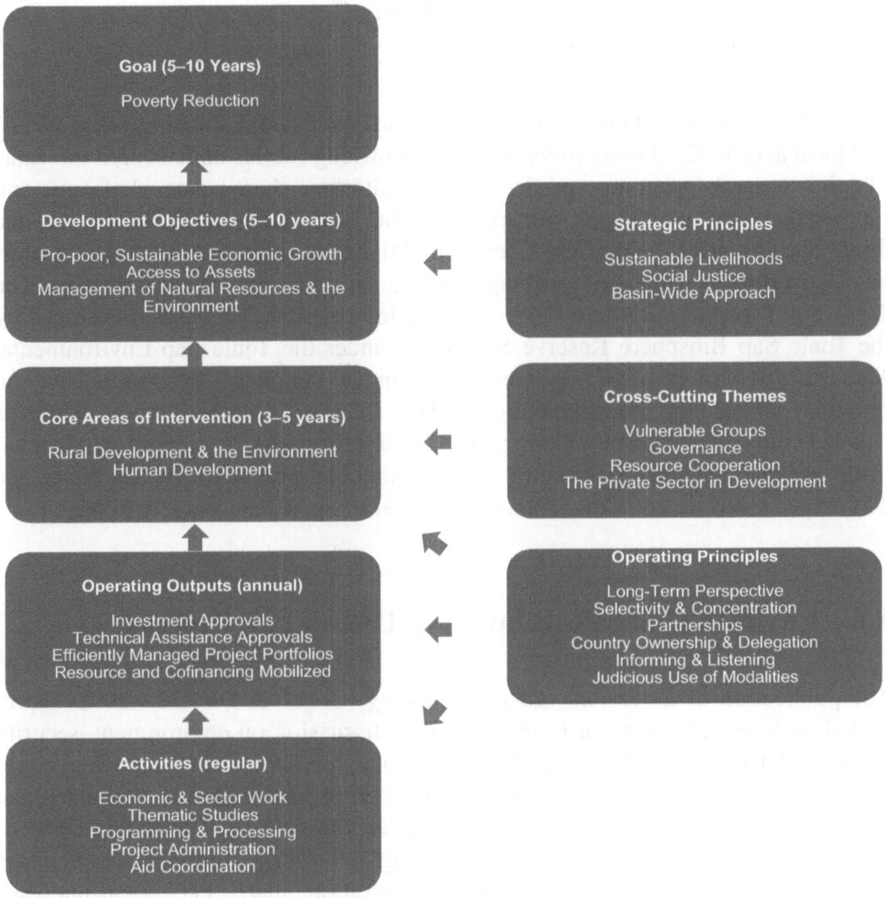

Fig. 8.1 The Tonle Sap Basin Strategy: Operating Framework. *Note*. From Serrat et al. (2005)

8.5.3 *Participatory Resource Management as Conflict Mitigation*

Beginning with the Tonle Sap Environmental Management Project (2002), the TSI placed strong emphasis on inclusive planning and community engagement. The *Tonle Sap Basin Strategy* (Serrat et al., 2005) identified community involvement as essential for legitimacy and sustainability. The 2005 document detailed how participatory processes were institutionalized through community fisheries committees, participatory land-use planning, community-based monitoring systems, and local conservation groups. These mechanisms reduced the potential for resource-related grievances and fostered trust among stakeholders, making collaborative governance a tool for both development and conflict mitigation.

8.5.4 Knowledge Platforms and Joint Problem-Solving Mechanisms

The TSI fostered shared knowledge systems that connected scientists, policymakers, and local actors. The *Future Solutions Now* knowledge series underscored the value of joint research programs, participatory monitoring, shared data platforms, and multi-stakeholder workshops. For example, the 2006 ADB–CAM Briefing on the Tonle Sap Initiative, held at the Council of Ministers, sought to strengthen senior-level government capacity to manage, engage in, and support the initiative. The Tonle Sap Biosphere Reserve Environmental Information Database—developed by the Tonle Sap Biosphere Reserve Secretariat under the Tonle Sap Environmental Management Project (2002)—served as a bilingual (English/Khmer) repository to promote sustainable management and data sharing. Together, these platforms enabled collective problem-solving and reduced information asymmetries, a mechanism later acknowledged as central to environmental peacebuilding. In effect, the TSI helped cultivate epistemic communities that bridged state–society divides.

8.6 Environmental Peacebuilding Theory and Practice

The publication of *Environmental Peacemaking* (Conca & Dabelko, 2002) represented an important moment in the emerging discussion on environment–security linkages. Their edited volume synthesized scholarship on how environmental cooperation could serve as a pathway to peace, offering a conceptual vocabulary that would shape the field over the next two decades. Specifically, Conca and Dabelko (2002) identified three core mechanisms through which environmental cooperation can contribute to peace: institutionalized cooperation, shared ecological interests, and confidence-building across political or social divides. These mechanisms were not presented as abstract theory but as empirically grounded pathways observed in transboundary water management, regional environmental agreements, and community-level natural resource governance. Their work built on earlier environmental security scholarship (Homer-Dixon, 1999; Matthew et al., 2002) but shifted the analytical lens from diagnosing risks to identifying cooperative opportunities—a shift the TSI was already operationalizing.

8.7 Mechanisms of Environmental Peacebuilding in the Tonle Sap Initiative

Viewed retrospectively, the alignment between Conca and Dabelko's (2002) core mechanisms and the governance architecture of the TSI is striking—especially given that the TSI was formulated and launched synchronously with the publication

Table 8.1 Alignment Between the TSI and Conca and Dabelko's (2002) Environmental Peacemaking Mechanisms

Environmental Peacemaking Mechanism (Conca & Dabelko, 2002)	Tonle Sap Initiative Mechanism
Institutionalized cooperation	Community fisheries; inter-ministerial coordination; multi-level governance architecture; Tonle Sap Authority; Tonle Sap Biosphere Reserve Secretariat; etc.
Shared ecological interests	Basin-wide framing; Tonle Sap Environmental Management Project (2002); Tonle Sap Sustainable Livelihoods Project (2005); advisory technical assistance for fisheries baseline assessments, co-management planning, GIS-based ecological monitoring, hydrological and sedimentation studies, watershed rehabilitation; etc.
Confidence-building	Community fisheries; local monitoring; multi-stakeholder workshops; participatory planning; participatory resource mapping; etc.
Addressing structural vulnerabilities (derived from Conca and Dabelko's discussion of contextual factors)	Environmental monitoring and enforcement capacity; governance strengthening; livelihood diversification; etc.

Note. This table illustrates how the TSI, launched in 2001, not only aligned with but operationalized these synthesized mechanisms of environmental peacemaking—(1) institutionalized cooperation, (2) shared ecological interests, and (3) confidence building. In addition, the TSI's governance and livelihood interventions addressed structural vulnerabilities that, while not presented by Conca and Dabelko (2002) as a discrete mechanism, are implicit in their treatment of contextual factors shaping environmental cooperation

of *Environmental Peacemaking*. Table 8.1 below shows how the TSI operationalized the mechanisms of environmental peacemaking as synthesized from Conca and Dabelko's (2002) broad conceptual discussion in Chap. 1 of the volume.

8.7.1 Institutionalized Cooperation

The TSI's governance architecture embodied the principle that environmental cooperation can create durable organizational structures that foster trust and reduce the likelihood of conflict. The *Tonle Sap Basin Strategy* (Serrat et al., 2005) called for cross-sectoral and cross-scalar coordination mechanisms, and this vision was operationalized through the Tonle Sap Biosphere Reserve Secretariat; inter-ministerial working groups; provincial coordination committees; community-based organizations; and partnerships with the Food and Agriculture Organization; the Government of Finland; the Global Environment Facility; national and international nongovernment organizations; the United Nations Development Programme; the United Nations Educational, Scientific and Cultural Organization, and the WorldFish

Center, among others. These arrangements were further elaborated in *From Strategy to Practice: The Tonle Sap Initiative* (Serrat & Moffatt, 2006), which emphasized the need for institutional innovation to address the region's socio-ecological challenges. By creating predictable, repeated interactions among actors who had previously operated in isolation, the TSI transformed fragmented governance into a cooperative regime—precisely the structural effect environmental peacebuilding seeks to generate.

8.7.2 Shared Ecological Interests

Environmental peacebuilding theory emphasizes that shared ecological dependencies can motivate collaboration even among actors with divergent interests. The Tonle Sap region's ecological interdependence—fisheries, hydrology, flood cycles—created a natural incentive for collaboration. The *Tonle Sap Basin Strategy* (Serrat et al., 2005) framed the lake as a shared ecological system whose sustainability depends on collective action. Between 2002–2006, this conceptualization was operationalized through:

- The Tonle Sap Environmental Management Project (2002)[5];
- The Tonle Sap Sustainable Livelihoods Project (2005)[6]; and
- Multiple ADB-financed advisory technical assistance initiatives—including support for strengthening the Inland Fisheries Research and Development Institute; the establishment of the Tonle Sap Basin Management Organization (a transitional, capacity-building mechanism whose work paved the way for the creation of the Tonle Sap Authority in 2007); capacity building for the TSI; a national environmental education and awareness campaign; improvements to the regulatory and management framework for inland fisheries; studies on the influence of built structures on the fisheries of the Tonle Sap; and efforts to enhance the role of women in inland fisheries.

These projects and advisory technical assistance were not isolated interventions. They formed an integrated basin-wide governance system that recognized the Tonle Sap as a shared ecological resource requiring joint management. Such linkages are

[5] The Tonle Sap Environmental Management Project (2002) comprised three interrelated components: (1) strengthening natural resource management coordination and planning for the Tonle Sap Biosphere Reserve; (2) organizing communities for natural resource management in the Tonle Sap Biosphere Reserve; and (3) building management capacity for biodiversity conservation in the Tonle Sap Biosphere Reserve.

[6] The Tonle Sap Sustainable Livelihoods Project (2005) consisted of three closely linked components: (1) supporting community-driven development; (2) safeguarding the three core areas of the Tonle Sap Biosphere Reserve—Boeng Chhmar, Prek Toal, and Stung Sen; and (3) building skills and awareness for sustainable livelihoods.

widely understood as essential for reducing mistrust and preventing the escalation of local disputes.

8.7.3 *Confidence-Building*

Conca and Dabelko (2002) argue that environmental cooperation can build confidence and reduce mistrust. The TSI's participatory planning processes brought together stakeholders who had historically operated in isolation or competition. *From Strategy to Practice: The Tonle Sap Initiative* (Serrat & Moffatt, 2006) underscored the role of participatory planning in fostering trust among communities, local authorities, national agencies, and nongovernment organizations. Participatory mechanisms included community fisheries committees, participatory land-use planning, community-based monitoring, local conservation groups, and multi-stakeholder workshops. These processes created horizontal linkages among communities and vertical linkages between communities and state institutions—hallmarks of environmental peacebuilding.

8.7.4 *Addressing Structural Vulnerabilities*

Environmental peacebuilding literature identifies structural vulnerabilities—poverty, weak governance, livelihood insecurity—as drivers of conflict. The TSI targeted these underlying risks directly. The Tonle Sap Environmental Management Project (2002) addressed fisheries governance, protected area management, environmental monitoring, and enforcement capacity. The Tonle Sap Sustainable Livelihoods Project (2005) focused on livelihood diversification, microfinance, community-based enterprise development, and women's economic empowerment. The *Future Solutions Now* knowledge series documented how these interventions reduced livelihood insecurity and strengthened governance capacity. By addressing the structural drivers of conflict, the TSI embodied the preventive logic of environmental peacebuilding, targeting not symptoms but root causes of potential instability.

8.8 Comparative and Theoretical Implications

The Tonle Sap Initiative offers a powerful case for rethinking the intellectual history of environmental peacebuilding. The initiative demonstrates that the practice of linking environmental cooperation to peace emerged before the concept was formally articulated in academic literature. This has important implications for how the field understands its origins, evolution, and future trajectory.

8.8.1 Revisiting the Practice–Theory Relationship

The prevailing account presents *Environmental Peacemaking* (Conca & Dabelko, 2002) as the field's main conceptual point of departure. While foundational, the TSI demonstrates that practitioners were already implementing integrated environmental governance approaches aligned with environmental peacebuilding logic. This suggests that development practice helped generate the conceptual foundations later codified by scholars, reversing the assumed direction of influence in the field's intellectual history.

8.8.2 Development Institutions as Early Innovators

Environmental peacebuilding scholarship has focused largely on transboundary water agreements, regional environmental cooperation, post-conflict resource governance, and peace parks. Far less attention has been given to development institutions as early innovators. The TSI shows that organizations such as the Asian Development Bank were designing integrated governance systems that addressed structural drivers of conflict.[7] Recognizing this broadens the field's genealogy and highlights development agencies as laboratories of institutional experimentation, challenging the assumption that environmental peacebuilding emerged primarily from academic or diplomatic circles.

8.8.3 Environmental Governance as Peace Infrastructure

Environmental peacebuilding literature increasingly views environmental governance as a form of peace infrastructure—institutions, norms, and practices that reduce conflict risk by addressing structural vulnerabilities and fostering cooperation. The TSI exemplifies this logic. Its governance architecture established predictable interactions among stakeholders; institutionalized cooperation across sectors and scales; strengthened institutions; improved transparency and accountability;

[7] Other early examples of projects and programs designed and implemented in fragile, conflict-affected, or post-conflict settings—and that addressed resource-related drivers of tension, institutional fragility, or cross-community cooperation—include African Development Bank watershed and rangeland programs in Somalia (2001–2004) and the Niger Basin (2002–2005); European Bank for Reconstruction and Development transboundary water-cooperation projects in Central Asia (2000–2004); Inter-American Development Bank watershed and sustainable-development initiatives in Haiti (2001–2004) and Colombia (2002–2005); Islamic Development Bank drought-resilience programs in the Sahel (2000–2005); and World Bank natural-resource projects in Afghanistan (2003) as well as water-management projects in the Great Lakes region (2001–2004), Sudan (2000–2005), and the West Bank and Gaza (2000; 2004).

reduced livelihood insecurity; and enhanced environmental monitoring and enforcement. Together, these functions show how environmental governance can serve as a durable foundation for peace, even without explicit peacebuilding mandates.

8.8.4 The Importance of Context-Specific Institutional Design

The TSI underscores the importance of context-specific approaches in environmental peacebuilding. Its effectiveness depended on understanding the Tonle Sap's socio-ecological dynamics, engaging local communities, and adapting governance structures to local realities. This grounded design strengthened environmental outcomes and social cohesion, demonstrating that environmental peacebuilding is most effective when embedded in lived realities rather than imposed through external templates.

While the Tonle Sap Initiative aligns closely with mechanisms later associated with environmental peacebuilding, it was never framed as a peacebuilding program. Its architects spoke the language of development, environmental management, and governance—not conflict prevention. Yet environmental peacebuilding is defined by mechanisms and effects, not by contemporary labels. By fostering cooperation, reducing livelihood insecurity, and strengthening institutions across scales, the TSI contributed to peace-enabling conditions now recognized as foundational. The absence of explicit peace rhetoric underscores how peacebuilding can emerge organically from integrated development practice.

A second counterargument concerns causality. Cambodia's political trajectory has remained fragile, and it would be inappropriate to claim that the TSI produced measurable national-level peace outcomes. The argument here is more modest and analytically precise: the TSI addressed specific conflict-risk factors—elite capture of fisheries, livelihood vulnerability, opaque governance, and weak enforcement—that environmental security scholars identify as potential drivers of tension. By clarifying rights, improving transparency, and institutionalizing participation, the TSI helped mitigate localized grievances. In this sense, it contributed to peace infrastructure not by transforming national politics but by strengthening everyday governance arrangements through which conflict risks are managed.

Finally, development interventions can themselves generate tensions, particularly when they alter access to resources or empower some actors over others. Decentralization, fisheries reforms, and protected area management are inherently political and can provoke resistance. The TSI was not immune to these dynamics.[8]

[8] To wit, the three core areas of the Tonle Sap Biosphere Reserve were intended to function as strictly protected conservation zones under the United Nations Educational, Scientific and Cultural Organization's Man and the Biosphere Programme and the Tonle Sap Environmental Management Project (2002). Yet their day-to-day management reflected the lake's fragmented institutional landscape. The Ministry of Environment held formal authority, but the Department of Fisheries (later the Fisheries Administration) retained a substantial operational role because of its long-standing

What distinguishes it is its deliberate emphasis on community-based institutions, iterative learning, and participatory mechanisms—an approach that sought to anticipate and mitigate distributional tensions rather than ignore them (Serrat, 2017a). This acknowledgement enhances analytical credibility and aligns the précis with contemporary understandings of conflict-sensitive development.

8.9 Legacy and Influence

The TSI informed Cambodia's fisheries reform process, protected area governance, decentralization policies, climate adaptation planning,[9] and rural development strategies. These contributions helped embed integrated environmental governance into national policy frameworks. Yet the TSI's influence extended beyond the Tonle Sap region. It contributed to regional environmental governance in Southeast Asia by demonstrating the feasibility of basin-wide systems; providing lessons for the Mekong River Commission (Serrat, 2025a)—particularly through the relocation of the Tonle Sap Biosphere Reserve Secretariat to the Cambodia National Mekong Committee in 2001[10]; informing regional climate-adaptation discussions; and supporting environmental cooperation initiatives within the Association of Southeast Asian Nations. Through these channels, the TSI helped shape regional approaches to environmental governance and resilience. Its influence thus spanned from local governance to national policy and regional cooperation, amplifying its peace-enabling effects.

---

enforcement mandate and the economic centrality of the lake's fisheries. This dual structure did not change the legal protected status of the core areas, but it did create overlapping mandates that produced uneven levels of protection across sites.

[9] These contributions draw on the accomplishments of ADB's Climate-Resilient Rice Commercialization Sector Development Program (2013)—which advanced drought- and flood-tolerant agricultural practices, climate-smart irrigation, and adaptive rural livelihoods—and those of the Integrated Urban Environmental Management in the Tonle Sap Basin Project (2015)—which strengthened flood-risk management, climate-resilient urban drainage, and solid-waste systems—alongside ADB's broader portfolio of climate-adaptive water resource interventions across the Tonle Sap region. ADB's Second Integrated Urban Environmental Management in the Tonle Sap Basin Project (2018) expanded climate-resilient urban infrastructure, enhanced flood protection, and improved adaptive water and wastewater management.

[10] The Tonle Sap basin is a critical sub-system of the wider Mekong basin, contributing a significant share of the system's seasonal water storage, ecological productivity, and basin-wide fish migrations. Its distinctive flood-pulse dynamics—driven by the annual reversal of the Tonle Sap River as rising Mekong River levels push water upstream—expand the lake and create extensive floodplain habitats. This tight hydrological and ecological interdependence means that governance decisions in the Tonle Sap basin, as well as fisheries and water-management choices made upstream in the Mekong mainstream and tributaries, have far-reaching implications for ecological stability, food security, and regional cooperation across the basin.

8.10 Rethinking Environmental Peacebuilding

The TSI offers a useful illustration of environmental peacebuilding in practice even as academic terminology was still taking shape. Reexamining the TSI through this lens expands the field's intellectual genealogy and highlights how peace-enabling conditions can emerge from development practice before they are formally named in scholarship. The TSI makes clear that practice and theory were evolving concurrently; development agencies were early innovators; environmental governance can function as peace infrastructure; and locally attuned institutional design is essential. It challenges the field to reconsider where environmental peacebuilding emerged and what work has been overlooked in its intellectual history.

Equally important, the TSI reveals that environmental peacebuilding is neither linear nor risk-free (Serrat, 2017b). It unfolds through experimentation, negotiation, and adaptation—often amid institutional fragility and political uncertainty. Acknowledging these complexities does not diminish the TSI's significance; it highlights the practical realities through which environmental cooperation contributes to stability. Environmental peacebuilding, as the TSI illustrates, evolves iteratively, adapts to shifting conditions, and is rooted in everyday governance practice.

The TSI's enduring imprint continues to shape environmental governance in Cambodia.[11] It also offers broader insights for the field: durable peace is often built not through grand diplomatic gestures but through the steady work of strengthening institutions, empowering communities, and managing shared ecological systems with transparency and care. Such lessons will be best harvested with evaluation frameworks designed to capture the conflict-sensitive, ecological, and governance dimensions of environmental peacebuilding—frameworks that may ultimately reshape how we understand peace itself (Serrat, 2025b). Box 8.1 presents my interview-based account of the TSI's design and early years, reproduced from *ADB: Reflections and Beyond* (ADB, 2009)—a publication I later conceived and oversaw while serving as principal knowledge management specialist and concurrently

[11] Other Tonle Sap-related projects approved by ADB after the Tonle Sap Environmental Management Project (2002) and the Tonle Sap Sustainable Livelihoods Project (2005), not listing ADB-financed advisory technical assistance, include the Tonle Sap Lowlands Rural Development Project (2006); the Tonle Sap Rural Water Supply and Sanitation Sector Project (2006); the Tonle Sap Poverty Reduction and Smallholder Development Project (2009), along with its additional financing (2015); the Second Rural Water Supply and Sanitation Sector Project (2009); the Water Resources Management Sector Development Program (2010); the Climate-Resilient Rice Commercialization Sector Development Program (2013); the Integrated Urban Environmental Management in the Tonle Sap Basin Project (2015); and the Second Integrated Urban Environmental Management in the Tonle Sap Basin Project (2018). Galvanized by the TSI, development partners have continued or increasingly directed their attention to the Tonle Sap basin and its wider watershed—including the Food and Agriculture Organization, ForumCiv, the Green Climate Fund, the International Fund for Agricultural Development, the United Nations Development Programme, and a range of philanthropic foundations—each contributing to climate-resilience, livelihood, and ecosystem-oriented programs in the region.

heading ADB's Knowledge Management Center, to capture the memories and experiences of its staff and to offer a more personal, human perspective on the challenges of development work.

Box 8.1 Failing Forward

Cambodia in my case was a pure accident. In 2001, I was working in ADB's European Representative Office in Frankfurt. A former Director of mine asked me to come back and help address a difficult situation that was rearing its head by the day—the Tonle Sap. I had never looked at a map of Cambodia; I had never heard of the name of the great lake, which is literally what Tonle Sap means, and so I googled the words and saw what appeared. I was taken aback: I had no idea that such a big lake existed; it is the largest freshwater lake in Southeast Asia, and a biosphere reserve of the United Nations Educational, Scientific, and Cultural Organization.

I had no notion that I would ever go to Cambodia. I carried on for 5 full years. But, between 2001 and the beginnings of ADB's operations we had had a decade of good work. You need to go back into history, and, of course, we do not want to refer to what occurred under the Khmer Rouge. After the Paris Accords and Cambodia joining ADB, basically from 1992/1993, ADB operations began full scale. ADB was not the only partner of the Government of Cambodia; there was great appetite and interest not just among multilateral development banks and bilateral agencies but also from nongovernment organizations, research institutes, and others, to help Cambodia. That provided a very unusual context for ADB in a country that was about to try to reconstruct itself, having undergone a terrible genocide—a sudden influx of agencies all meaning the best for the country but, of course, putting pressure on rather depleted capacities.

In the early stages, because of the havoc caused by the Khmer Rouge, everything was destroyed—not just the people as a living entity but the infrastructure was also completely demolished. And so, much of the first efforts of agencies such as ADB were very much into building infrastructure. I understand that in the early years of ADB involvement, there was a strong emphasis on the transport sector. That was a shrewd move because rebuilding networks not only for trade but also to reestablish lines of communication is important. The manner in which ADB and a few other agencies decided to go about it was quite interesting. They saw that there was a population that needed to be employed, so instead of giving them top-range technologies, they decided to construct laterite roads. These would be literally built by hand by the communities themselves, providing gainful employment as well as infrastructure.

(continued)

Box 8.1 (continued)

ADB realized also that agriculture, rural development, and natural resources were areas worthy of interest, and this is probably how I came into the picture with the Tonle Sap Initiative. Originally, ADB's interest in the Tonle Sap was quite ahead of its time because nobody was doing much there when ADB began, except perhaps the Food and Agriculture Organization through a very small pilot project that went on for many years.

When I arrived for the first time in Cambodia, I naturally went to a great many meetings. One has to enjoy these when one works in such organizations. So I spent the first days in something like 10–12 meetings a day, coming out with absolutely roaring headaches. On the fourth day, I said, "I can't take this anymore. There is too much information, I must see the lake." So I fled Phnom Penh and took a car to Siem Reap.

The roads were pretty much gone and I slipped a disc on the way. The trip took something like 9 h. As I arrived, just overlooking the hills coming into Siem Reap, the rain began to fall and, for some reason, rain has always played an important part in my mental processes: it irrigates them. I had become steadily more depressed during the 9-h trip from Phnom Penh, questioning the wisdom of trying to do something on the Tonle Sap through one regional technical assistance. By the time I arrived in Siem Reap, I said, "There is no project. It is too complicated." Then logic kicked in. I said, "Well, if there is no project, what is there?" The answer came: "There are projects."

In hindsight it sounds simple, but it was not easy to come to that conclusion knowing that if you talk in terms of projects you have to leverage extensive resources. You have to come up with a strategic framework against which to couch various interventions in partnerships with a number of other agents. But I felt confident, for whatever reason, that this was the right way forward, and then proposed to conceptualize the Tonle Sap Basin Strategy. It promotes a long-term, holistic approach that conserves nature and offers the promise of sustainable development.

The major components of complexity are clearly related to the hydrology of the lake which, as you may know, reverses itself. During the rainy season, basins normally discharge into the Mekong River. But, in Cambodia, the power of the Mekong River pushes the Tonle Sap River which drains the great lake in the opposite direction. Instead of emptying itself, the lake fills covering four or five times its original dry season area, therefore flooding a lot of the ground around the lake. This means that people around the lake live in a highly unusual environment. Sometimes they have to construct floating pig pens, or build two-storey houses. Sometimes they travel by moped, other times by boat. They are, in turn, agriculturalists and fishers.

(continued)

Box 8.1 (continued)

A lot of the infrastructure that you use during the dry season you cannot use during the wet season. The roads that remain are washed over by the lake which increases from something like one m in depth at its lowest up to 12 m. It is the reverse hydrology of the lake that changes the livelihoods of people around the lake. This happens over a substantial area because as I said the lake increases by up to five times its surface area. It is this harmonious complexity that requires a bit of thinking and a bit of understanding.

We had to start somewhere. The more complex issues we thought were centered around the lake, so that provided the entry point for the first project which we entitled Tonle Sap Environmental Management. That was a small loan of $10 million. We kept it as small as possible because Cambodia is a small country that needs to repay these loans. Thankfully, the Global Environment Facility extended $4 million in grant form. The United Nations Development Programme also chipped in $600,000. A nice package.

We were quite ambitious when you look at the reforms. We wanted to be encouraged by the state's response. If it was ready to take the bull by the horns, this would in a sense validate further investments. So we centered the first project on the lake itself—looking at matters such as the fisheries policy of the country, the core areas that are meant to be sacrosanct protected areas, mapping of the boundaries of community fisheries, and so on.

Having endeavored to provide a sustainable policy and institutional environment, we felt that the time would then come to deliver investment packages as directly as possible into the hands of the communities, hence, the next stage of the strategy: Tonle Sap Sustainable Livelihoods. This moved us away from the lake because, of course, communities do not live in the center of it. They are on its peripheries. The next project we planned was originally titled Tonle Sap Lowland Stabilization—I think it ended up being called Tonle Sap Lowlands Rural Development. This was to be more of the same, following on from the model established under Tonle Sap Sustainable Livelihoods: delivering quick packages, sector program interventions involving irrigation and water supply, etc. into the lowland areas that equate with the transition area of the biosphere reserve. That was the third pillar of the Tonle Sap Basin Strategy. The fourth one would tackle deforestation in the watersheds and this I understand was to be prepared sometime this year. The phasing of the projects ensures that each is informed by and feeds another, promoting continuity and building synergies. It is envisaged that most elements of the cycle will go through several iterations. Much advisory technical assistance, often small-scale, would instruct and synergize the core projects, to which others are invited to ally themselves in partnerships.

(continued)

Box 8.1 (continued)

We knew that we would fail on the way because of the complexity of development work in general, but we wanted to "fail forward," as a colleague of mine terms it. Failing forward means that you capture the lessons of experience and you endeavor not to repeat mistakes.

The going got rough but I think the first project was the most difficult to formulate as it was the father of the series. In ADB, questions were asked about what would come next and whether we could pull it off. It was the same with the government. Cambodia was willing to borrow $10 million, for a project that comprised little but software. We were building institutions and we were helping them reform policies and put in place strategies and capacities. For a small country, agreeing to $10 million in the form of such a loan is remarkable. (ADB, 2009, pp. 72–76).

References

ADB. (2009). Failing forward. In: Asian Development Bank (Ed.), *ADB: Reflections and beyond* (pp. 72–76). https://www.adb.org/publications/adb-reflections-and-beyond

Conca, K., & Dabelko, G. (Eds.). (2002). *Environmental peacemaking*. Johns Hopkins University Press.

Homer-Dixon, T. (1999). *Environment, scarcity, and violence*. Princeton University Press.

Matthew, R., Halle, M., & Switzer, J. (Eds.). (2002). *Conserving the peace: Resources, livelihoods, and security*. International Institute for Sustainable Development.

Serrat, O. (2002, October). *Future Solutions Now: The Tonle Sap Initiative*. Asian Development Bank. https://www.researchgate.net/publication/268516589_Future_Solutions_Now_-_The_Tonle_Sap_Initiative

Serrat, O. (2003a, April). *Future Solutions Now: The Tonle Sap Initiative*. Asian Development Bank. https://www.researchgate.net/publication/268516750_Future_Solutions_Now_-_The_Tonle_Sap_Initiative.

Serrat, O. (2003b, August). *Future Solutions Now: The Tonle Sap Initiative*. Asian Development Bank. https://www.researchgate.net/publication/268516595_Future_Solutions_Now_-_The_Tonle_Sap_Initiative.

Serrat, O. (2003c, December). *Future Solutions Now: The Tonle Sap Initiative*. Asian Development Bank. https://www.researchgate.net/publication/268516751_Future_Solutions_Now_-_The_Tonle_Sap_Initiative.

Serrat, O. (2004a, November). *Future Solutions Now: The Tonle Sap Initiative*. Asian Development Bank. https://www.researchgate.net/publication/268516754_Future_Solutions_Now_-_The_Tonle_Sap_Initiative.

Serrat, O. (2004b, December). *Future Solutions Now: The Tonle Sap Initiative*. Asian Development Bank. https://www.researchgate.net/publication/268516603_Future_Solutions_Now_-_The_Tonle_Sap_Initiative.

Serrat, O. (2005, November). *Future Solutions Now: The Tonle Sap Initiative*. Asian Development Bank. https://www.researchgate.net/publication/268516605_Future_Solutions_Now_-_The_Tonle_Sap_Initiative.

Serrat, O. (2006, May). *Future Solutions Now: The Tonle Sap Initiative*. Asian Development Bank. https://www.researchgate.net/publication/268516761_Future_Solutions_Now_-_The_Tonle_Sap_Initiative.

Serrat, O., Gallego-Lizon, T., & Moffatt, D. (2005). *The Tonle Sap Basin Strategy*. Asian Development Bank. https://www.adb.org/documents/tonle-sap-basin-strategy

Serrat, O., & Moffatt, D. (2006). *From strategy to practice: The Tonle Sap Initiative*. Asian Development Bank. https://www.researchgate.net/publication/267242989_From_Strategy_to_Practice_The_Tonle_Sap_Initiative

Serrat, O. (2017a). From strategy to practice. In *Knowledge solutions: Tools, methods, and approaches to drive organizational performance* (pp. 107–110). Springer.

Serrat, O. (2017b). Understanding complexity. In *Knowledge solutions: Tools, methods, and approaches to drive organizational performance* (pp. 345–353). Springer.

Serrat, O. (2025a). Safeguarding the mother of waters: Environmental peacebuilding in the Mekong River Basin. In *Anthropogenic solutions for climate change: Achieving environmental peace* (pp. 229–239). Springer.

Serrat, O. (2025b). *Assessing environmental peacebuilding: Principles, criteria, and pathways for proactive interventions*. Unpublished manuscript, Georgetown University. https://www.researchgate.net/publication/398543700_Assessing_Environmental_Peacebuilding_Principles_Criteria_and_Pathways_for_Proactive_Interventions

Serrat, O. (2026). *Roots of reconciliation: The emergence of environmental peacebuilding*. Unpublished manuscript, Georgetown University. https://www.researchgate.net/publication/400080708_Roots_of_Reconciliation_The_Emergence_of_Environmental_Peacebuilding

Chapter 9
Environmental Peacebuilding: Futures Made Together

Abstract This précis offers a forward-looking synthesis of this SpringerBrief's central themes. It distills the Brief's insights into a reflection on how environmental peacebuilding must evolve in the Anthropocene. It highlights the shift from cooperation to co-creation, the growing importance of adaptive and context-sensitive evaluation, and the ethical and intergenerational stakes that will shape the field's next generation. In doing so, the précis frames environmental peacebuilding as an essential architecture for building shared futures in an increasingly interdependent world.

9.1 Foundations and Emerging Pressures

"In nature nothing exists alone," observed Rachel Carson. That recognition captures the spirit of environmental peacebuilding. Environmental peacebuilding has always been a field shaped by necessity. It arose from the recognition that environmental pressures do not wait for political settlements, and that ecological interdependence can draw societies together even when history pulls them apart. As the chapters of this SpringerBrief have shown, the field's foundations stretch across centuries of practice, decades of scholarship, and the lived experience of communities who have long understood that peace is inseparable from the landscapes that sustain it. The field did not arise from theory alone; it grew through the persistent work of communities who learned to negotiate access, share resources, and steward ecosystems long before the term "environmental peacebuilding" existed.

The world coming into view today is differs substantially from the one in which the field first developed two decades ago. Climate disruption is accelerating, ecological thresholds are being crossed, and cascading stresses—water scarcity, food insecurity, and climate-driven mobility—are reshaping geopolitical realities. These pressures are occurring as humanity approaches or exceeds multiple planetary boundaries, underscoring the urgency of governance approaches that recognize biophysical limits. Shifting energy and resource dynamics are amplifying these realignments, generating new dependencies even as they expose fresh fault lines.

O. Serrat, *Advancing Environmental Peacebuilding*, SpringerBriefs in Political Science, https://doi.org/10.1007/978-981-92-4028-9_9

Governance systems designed for a more stable era are straining under the weight of interconnected risks they were not built to manage. In this context, environmental peacebuilding is no longer a promising niche; it is becoming a necessary architecture for navigating the Anthropocene. Even so, meeting the pressures of this century requires a field capable of matching their scale, complexity, and speed. These transformations do more than challenge existing institutions—they compel societies to rethink what cooperation means and how it must be practiced in an era defined by shared vulnerability.

9.2 Co-Creation and the Expanding Landscape of Actors

Environmental peacebuilding's next phase will require a shift in orientation. Cooperation over shared resources will remain essential, but it will not be sufficient. This is where the field must evolve from cooperation to co-creation—the joint design of institutions, knowledge systems, and adaptive capacities that allow societies to live with uncertainty rather than be undone by it. Co-creation is not a method; it is a mindset. It requires deeper engagement with Indigenous governance, local knowledge, and the lived experience of communities on the front lines of environmental change. It requires institutions to listen as much as they lead and to recognize that resilience is built not only through technical interventions but through relationships of trust, reciprocity, and shared purpose.

Co-creation also means reimagining who participates in environmental peacebuilding. Women's cooperatives, Indigenous networks, youth movements, and transnational civil society organizations are increasingly shaping environmental governance in ways that states alone cannot. Their contributions are not ancillary; they are central to the legitimacy and durability of environmental cooperation. Environmental peacebuilding, as it moves into its next phase, will need to embrace these actors not as stakeholders to be consulted, but as partners in designing the futures they will inherit.

9.3 Tools, Power, and Governance in a Changing Climate

New tools will shape this future. Advances in ecological monitoring, remote sensing, and artificial intelligence are transforming how environmental change is understood and acted upon. Satellite-based hydrological mapping can now detect shifts in river basins in near-real time. Machine-learning models can anticipate drought patterns months in advance. Community-based digital platforms can document environmental harms and governance failures with unprecedented speed. These technologies can reveal patterns that were once invisible, but they also raise questions about equity, access, and power. Who controls the data? Who interprets it? Who benefits from it? The challenge ahead is not simply to use new tools, but to embed them within governance systems that are transparent, accountable, and inclusive.

Yet even the most advanced tools cannot insulate societies from the geopolitical transformations already underway. The geopolitical implications of climate disruption will further shape the field. As sea levels rise, storms intensify, and agricultural systems destabilize, climate-driven mobility will become one of the defining features of the century. Regions already facing fragility—such as the Sahel, the Horn of Africa, and parts of South and Southeast Asia—will experience overlapping pressures that blur the boundaries between humanitarian response, development, and peacebuilding. Environmental peacebuilding will need to operate across these boundaries, helping societies navigate the tensions that arise when environmental change intersects with identity, sovereignty, and historical grievance. It will also need to support regional and transboundary governance arrangements capable of managing shared risks that no state can address alone.

9.4 Learning and Evaluation in Socio-Ecological Systems

Most of all, the field must strengthen its capacity to learn. Without evaluation frameworks capable of capturing socio-ecological complexity, environmental peacebuilding risks becoming a set of hopeful experiments rather than a coherent practice. The evaluative criteria proposed in this Brief—adaptability and resilience, conflict sensitivity and "do no harm", cultural sensitivity, inclusivity and participation, legitimacy, long-term environmental impact, and "peace writ large"—are offered as a starting point. They are not definitive. They invite us to build an evaluative architecture that matches the complexity of the systems we seek to understand and the stakes of the century we have entered.

Evaluation is not an administrative exercise; it is a form of governance. It shapes what is valued, what is measured, and what is learned, echoing arguments for learning-oriented evaluation systems (Serrat, 2021). In fragile contexts, where interventions unfold amid uncertainty and shifting baselines, evaluation must be iterative, context-sensitive, and attuned to the lived realities of communities. It must capture not only whether cooperation occurred but how it was built, who benefited, who was excluded, and whether the ecological foundations of peace were strengthened or eroded. Without such evaluative clarity, environmental peacebuilding cannot mature into a field capable of guiding policy, informing investment, or shaping institutional practice.

9.5 Ethics, Responsibility, and the Work of Building Shared Futures

The ethical stakes of environmental peacebuilding will deepen further. The Anthropocene forces us to confront questions of intergenerational responsibility, ecological limits, and the moral dimensions of shared vulnerability. Environmental peacebuilding cannot be content with managing scarcity; it must help societies imagine and enact forms of stewardship that honor the rights of future generations. It must grapple with the uneven distribution of environmental harms and benefits, and with the historical injustices that shape contemporary vulnerabilities. It must ensure that environmental cooperation does not mask inequities or entrench power asymmetries, but instead contributes to more just and inclusive forms of governance.

Environmental peacebuilding's promise lies not in its ability to prevent all conflict—no field can do that—but in its capacity to cultivate the relationships, institutions, and ecological conditions that make peace more likely and more durable. Its future depends on whether we can move from reacting to crises to anticipating them, from managing scarcity to stewarding resilience, and from negotiating interests to imagining shared futures. The work ahead is demanding. But it is also generative. It invites us to rethink how societies relate to one another and to the natural systems that sustain them.

Environmental peacebuilding reminds us that cooperation is a practical, context-specific process grounded in how societies manage shared resources and ecological systems. It is built in the restoration of degraded landscapes, in the negotiation of shared water, in the labor of communities rebuilding trust after conflict, and in the collective recognition that our futures are intertwined. In a twenty-first century defined by interdependence, this may be one of the most important forms of peace we can build. In the end, environmental peacebuilding asks us to choose care over depletion and shared purpose over short-term gain.

9.6 Toward a Planetary Evaluation Ethic

The future will not be inherited—it will be made together. The supplementary evaluative criteria proposed in this Brief are meant to move environmental peacebuilding beyond the constraints of the OECD–DAC framework. Yet even these advances must now be situated within a broader evaluative horizon. Patton's (2020) *Blue Marble Evaluation* offers precisely such a horizon, articulating four overarching principles that anchor evaluation in a globally interdependent world: (1) Global Thinking, which insists that evaluators attend to cross-scale interconnections and planetary systems; (2) Anthropocene as Context, which situates all evaluation within the realities of global environmental change and biophysical limits; (3) Transformative Engagement, which orients evaluation toward systems transformation rather than incremental improvement; and (4) Integration, which weaves together diverse knowledge systems, disciplines, and perspectives to illuminate complex socio-ecological dynamics. These principles resonate with the trajectory traced in this Brief. They extend the field's logic from cooperation to co-creation, from context-sensitive evaluation to planetary stewardship, and from project-bounded learning to an evaluative practice capable of illuminating how environmental governance contributes to peace across scales. Integrating Blue Marble principles into environmental peacebuilding evaluation would strengthen the field's capacity to anticipate rather than merely react, to steward resilience rather than manage scarcity, and to align environmental peacebuilding with the ethical and intergenerational responsibilities of the Anthropocene. In this sense, *Blue Marble Evaluation* (Patton, 2020) does not replace the supplementary evaluative criteria proposed here; it situates them within a larger architecture of planetary responsibility. This shift demands not only new ways of knowing but new commitments to acting on that knowledge. "Knowing is not enough; we must apply. Willing is not enough; we must do."—Johann Wolfgang von Goethe.

References

Patton, M. (2020). *Blue marble evaluation: Premises and principles*. Guilford Press.

Serrat, O. (2021). Change management strategy in action: Independent evaluation for learning. In *Leading solutions: Essays in business psychology* (pp. 295–303). Springer.

Index

O. Serrat, *Advancing Environmental Peacebuilding*, SpringerBriefs in Political Science, https://doi.org/10.1007/978-981-92-4028-9

MIX
Papier aus verantwortungsvollen Quellen
Paper from responsible sources
FSC® C105338

If you have any concerns about our products,
you can contact us on
ProductSafety@springernature.com

In case Publisher is established outside the EU,
the EU authorized representative is:
Springer Nature Customer Service Center GmbH
Europaplatz 3, 69115 Heidelberg, Germany

Printed by Libri Plureos GmbH
in Hamburg, Germany